Human iPS Cells in Disease Modelling

Keiichi Fukuda
Editor

Human iPS Cells in Disease Modelling

Springer

Editor
Keiichi Fukuda
Department of Cardiology
Keio University School of Medicine
Tokyo, Japan

ISBN 978-4-431-55964-1 ISBN 978-4-431-55966-5 (eBook)
DOI 10.1007/978-4-431-55966-5

Library of Congress Control Number: 2016936099

Printed on acid-free paper

This Springer imprint is published by Springer Nature
The registered company is Springer Japan KK

Preface

Recent advances in stem cell biology have enabled us to examine the regeneration of various tissues and organs. One of the biggest advances in this field is the induced pluripotent stem cell (iPS cell), which was developed by Shinya Yamanaka in 2006. The iPS cell was a tailor-made multipotent stem cell, and was generated by transfection of the combination of several embryonic stem (ES) cell-specific transcription factors such as Oct3/4, Sox2, and Klf4. It has pluripotency, and can differentiate into various types of cells such as ES cells. Because the iPS cells maintained all the genome information including HLA, it cannot be immune-rejected when its derived cells are transplanted to the host. In 2014, the first clinical trial was performed by Masayo Takahashi in patients with senile macular degeneration using autologous iPS cell-derived retinal pigmented cells. Other clinical trials are now being conducted in diverse ways including the cornea, Parkinson's Disease, spinal cord injury, platelet production, and severe congestive heart failure. Realization of these projects has been eagerly awaited by patients with severe intractable diseases.

It was also expected that iPS cells could be used in another field: disease modelling. Hereditary diseases are caused by genome mutations, but their clinical phenotypes, severity, onset, and treatment show wide variation. Mouse models of human disease have been generated by gene targeting and transgenic animals during the past 25 years. Nevertheless, these animal models cannot always help us to simulate human disease and screening of drugs. Genes and proteins are different between mice and humans. Because human tissues and cells were not usually available in the in vitro experiments except for small amounts of tissue obtained by biopsy or autopsy, this became a big hurdle for in vitro phenotype analysis and drug development. Moreover, disease modelling of human cells was not available in a routine clinical and experimental setting. Generation of patient-derived iPS cells and induction of in vitro differentiation into the targeted cells and tissues greatly changed the situation. Use of iPS cells for the investigation of disease modelling and drug screening is certain to change the future direction of research and industry.

In this book, we asked the top scientists in the field to write about human iPS cells for disease modelling. This will greatly help readers to understand what an iPS cell is, how to make iPS cells from blood cells, how to apply these techniques to approaching a pathophysiological analysis, and how to perform drug development for patients with hereditary diseases. I strongly hope that readers can easily understand this field and will attempt disease modelling and tailor-made drug development for patients around the world.

Tokyo, Japan Keiichi Fukuda

Contents

Recent Improvements and Emerging Issues in iPSC Generation for the Modeling of Disease

1

Tomohisa Seki, Shinsuke Yuasa, and Keiichi Fukuda

Abstract

Recently, induced pluripotent stem cells (iPSCs) have attracted attention as a novel tool for the modeling of disease because of their potential to reveal new insights that have not been elucidated using animal models. Since iPSC generation was first reported, there have been many efforts to improve the method of generating iPSCs for clinical applications. To date, many methods for iPSC generation have been reported. Each has advantages and disadvantages for the modeling of disease, and thus the most appropriate method differs depending on the intended use of the iPSCs. Additionally, as the study of disease modeling with human iPSCs has progressed, the need to remove uncertainties due to variations in iPSCs cell lines has increasingly focused researchers' attention on attaining experimental accuracy. Recognition of these uncertainties is important for the advancement of disease modeling studies with iPSCs.

Keywords

Induced pluripotent stem cells • Disease modeling • Cell reprogramming

1.1 Introduction

Since the reprogramming of somatic cells by forced expression of reprogramming factors was first reported (Takahashi and Yamanaka 2006), this technique has attracted attention as a novel tool for regenerative therapy and disease research. The cells obtained using this approach have been named induced pluripotent stem

T. Seki • S. Yuasa • K. Fukuda, M.D., Ph.D., F.A.C.C. (✉)
Department of Cardiology, Keio University School of Medicine, 35 Shinanomachi, Shinjuku-ku, Tokyo 160-8582, Japan
e-mail: kfukuda@a2.keio.jp

K. Fukuda (ed.), *Human iPS Cells in Disease Modelling*,
DOI 10.1007/978-4-431-55966-5_1

cells (iPSCs). The first method for generating human iPSCs was based on retrovirally mediated introduction of genes into primary cultures of fibroblasts (Takahashi et al. 2007). Since then, many efforts have aimed at improving the generation of iPSCs for clinical applications. However, the best method for generating iPSCs is a topic for discussion. Each method has advantages and disadvantages, and the most preferable choice should be based on the intended use of the iPSCs.

The reason for generating iPSCs falls into one of two major classifications. One is for transplantation therapy (Garber 2013) and the other is for disease research. iPSCs generated from patients are not only a source of cells for transplantation but are also a model of human disease. In vitro modeling of disease with a patient's own cells can possibly lead to novel insights that would be unattainable using an animal model.

To date, there have been many efforts to improve methods for generating iPSCs for clinical applications. For example, establishing xeno-free culture conditions and selecting non-tumorigenic iPSC lines are becoming more important for the use of iPSCs in transplantation therapy (Lee et al. 2013). It must be emphasized that generation of iPSCs for disease modeling is different to that of cells for future transplantation therapy. Establishing noninvasive cell sampling methods from patients for generating iPSCs (Yamanaka 2010) and avoiding host genome anomalies in the iPSCs are particularly important points to consider for generating iPSCs for disease modeling. Additionally, it is possible that iPSC lines do not always have the same characteristics, even when generated from the same donor (Kajiwara et al. 2012). Therefore, to understand disease modeling using iPSCs, it is important to focus on the features of iPSC generation methods and the quality of iPSC lines.

1.2 Combination of Transgenes for Somatic Cell Reprogramming

In the first report of iPSC generation, forced expression of *OCT3/4*, *SOX2*, *KLF4*, and *C-MYC* was used for successful somatic reprogramming (Takahashi and Yamanaka 2006). Since then, the combination of reprogramming factors required to generate iPSCs has been one of the most hotly debated issues in iPSC research. In many situations, the quality of iPSCs has been seen as an important condition for clinical use. For instance, *C-MYC* is an oncogenic gene (Dang 2012). Following the first report of iPSC generation, *C-MYC* was shown to be dispensable for iPSC generation in the mouse, although the reprogramming efficiency was dramatically lowered with the combination of only three factors: *OCT3/4*, *SOX2*, and *KLF4* (Nakagawa et al. 2008). Subsequently, *TBX3* (Han et al. 2010), *L-MYC* (Nakagawa et al. 2010), or *Glis1* (Maekawa et al. 2011) were shown to serve as a substitute for *C-MYC* and have a salutary effect on the efficiency in reprogramming somatic cells or germ line transmission of generated iPSCs. But, of course, validation of iPSC quality such as by the efficiency of germ line transmission is confined to experiments with nonhuman iPSCs. Therefore, the combination of reprogramming factors appropriate for generating human iPSCs for modeling human disease has been a topic of discussion.

For generating iPSCs from patients, the efficiency of cell reprogramming has to be considered important because the chances of sampling a patient's cells or tissues are limited. Additionally, the amount of available tissue for generating a patient's iPSCs will be also limited. Therefore, establishing a stable system for generating human iPSCs forms the basis of experiments that model disease with iPSCs.

To date, many pathways have been recognized as targets for improving reprogramming efficiency. In particular, suppression of the p53-p21 pathway of tumor suppressor signals (Green and Kroemer 2009) enhances the efficiency of cell reprogramming (Hong et al. 2009; Kawamura et al. 2009; Li et al. 2009; Marion et al. 2009; Utikal et al. 2009). In fact, the short hairpin RNA (shRNA) of *p53* is included in the set of reprogramming factors used in the protocols provided by the Center for iPS Cell Research and Application (CiRA) to knock down the expression of *p53* and maintain the efficiency of somatic cell reprogramming (Okita et al. 2013). For establishing stable and efficient techniques for generating human iPSCs, the addition of factors that influence desired pathways and improve cell reprogramming efficiency is of considerable use.

1.3 Residual Transgene-Free Methods of Generating iPSCs

Initially, iPSCs were generated using a combination of fibroblast cultures and retroviral gene introduction (Takahashi et al. 2007; Takahashi and Yamanaka 2006). Subsequently, tumorigenic risk derived from reactivation of transgenes was reported (Okita et al. 2007), although transgenes that were introduced with retroviral vectors were silenced in the pluripotent state (Hawley et al. 1994; Stewart et al. 1982). Additionally, besides the influence of residual transgene expression, genomic insertion in itself has the potential to disrupt gene function and change gene expression in iPSCs. Interestingly, introduction of green fluorescent protein alone using lentivirus at an extremely high multiplicity of infection led to somatic cell reprogramming (Kane et al. 2010). This phenomenon is thought to be caused by multiple insertions of vector DNA into the genome and is also thought to be possible evidence for a harmful effect of insertional vectors. In cases of disease modeling with patient's iPSCs, factors that possibly affect the phenotype of patient-derived cells should be removed by all means. Therefore, gene introduction methods accompanied by genomic insertion are unfavorable for generating iPSCs, not only for transplantation therapy but also for disease modeling.

To avoid genomic insertion of transgenes, many methods have been established. Adenovirus (Stadtfeld et al. 2008), *Sendai virus* (Fusaki et al. 2009), transposons (Woltjen et al. 2009), RNA (Warren et al. 2010), recombinant protein (Kim et al. 2009; Zhou et al. 2009), and episomal vectors (Okita et al. 2011) are successful methods for introducing reprogramming factors and generating transgene-free iPSCs. However, these methods need additional checks to confirm the disappearance of transgenes in iPSCs. For instance, episomal vectors can potentially insert into the host genome at an extremely low rate. Although *Sendai virus* is a negative-strand virus and it does not integrate into host DNA in the course of gene expression, the elimination of *Sendai virus* from generated iPSCs should be confirmed by quantitative PCR. In the case of using transposons for generating iPSCs, additional treatment with transposase is required to eliminate the transposon sequence from the host genome. After the transposase treatment, additional checks are required to confirm the disappearance of the inserted sequence in the host genome. Therefore, additional robust evaluation of residual transgenes in iPSCs and accurate iPSC cell line selection are necessary for obtaining transgene-free iPSCs.

1.4 Noninvasive Procedure of Cell Sampling for iPSC Generation

Naturally, the purpose of disease modeling with iPSCs is to clarify the disease mechanism, to establish novel therapies, and, finally, to suit the interests of patients. Disease modeling with iPSCs is not invariably in the interests of patients who provide their own cells or tissues for generating iPSCs. Additionally, invasive methods of cell sampling may lead to a reduced number of donors for disease modeling. In cases of familial inherited disease, evaluation of the phenotypic differences between iPSC-derived cells generated from different family members is important for the reproducibility of results and assessment of the etiology. Therefore, focus should be on noninvasive cell sampling for iPSC generation when modeling disease, potentially more so than when using iPSCs for transplantation therapy.

As mentioned above, in the first report of iPSC generation, dermal fibroblasts were used as material to generate human iPSCs (Takahashi et al. 2007). To obtain patients' dermal fibroblasts, a dermal biopsy was needed. This procedure leaves a small scar on the patient's skin. To minimize the invasiveness of iPSC generation, many other less invasive methods to harvest patient-derived cell types have been reported: keratinocytes from plucked hair (Aasen et al. 2008), fibroblasts from oral mucosa (Miyoshi et al. 2010), and peripheral blood cells obtained by venipuncture (Loh et al. 2010; Seki et al. 2010; Staerk et al. 2010). Because of the ease in cell handling and universality of the procedure, generation of iPSCs from blood cells is becoming the preferred method. In fact, the official website of the CiRA currently provides protocols for generating human iPSCs from peripheral blood.

1.5 Genomic Abnormality in iPSCs

Even if the disappearance of transgenes in iPSCs is appropriately confirmed, researchers who use iPSCs for disease modeling should keep in mind that the iPSC genome potentially contains varied abnormalities. In the past, a high incidence of chromosomal aberrations was found in human iPSCs as copy-number variations (Hussein et al. 2011). Additionally, human iPSCs may also contain a high incidence of point mutations in protein-coding regions of the genome (Gore et al. 2011). iPSCs pose a potential risk in that they conserve somatic mutations from their cells of origin. In fact, rearrangements of genomic DNA, such as T cell receptor (TCR) genes in T cells (Nishimura et al. 2013; Vizcardo et al. 2013) and immunoglobulin (Ig) genes in B cells (Choi et al. 2011), remain in the genome after forced expression of reprogramming factors. Cell lines are cloned in the process of generating iPSCs; thus, the risk is that somatic chromosomal aberrations and mutations remain in the iPSCs.

Additionally, long-term culture of embryonic stem cells (ESCs) is known to result in chromosomal abnormalities (Baker et al. 2007). The circumstances in artificial culture encourage DNA damage in cells. Whether somatic cell reprogramming itself, through forced expression of transgenes, is mutagenic or not is an unresolved issue. The risk of a potential genomic abnormality in iPSCs has to be considered when modeling disease in these cells. Therefore, for correct interpretation of experimental results of disease modeling with iPSCs, it is necessary to appropriately select iPSC lines and reproduce results in several iPSC lines.

1.6 Differences Between iPSC Lines

Because of the possible genomic abnormalities in iPSC lines that originate from the host genome, the influence that these mutations have in the host may extend to the phenotypes of the iPSC cell lines. Additionally, epigenetic status also affects the character of iPSCs. For example, there are differences in the proclivities of teratomas formed from mouse iPSC-derived neural progenitor cells differentiated from several types of somatic cells (Miura et al. 2009). Additionally, iPSCs generated from different types of somatic cells have differentiation tendencies toward that of the donor cell type (Kim et al. 2010; Polo et al. 2010). As well as the iPSC differentiation tendencies, residual epigenetic features derived from the original cells also existed in the iPSCs described in these reports. Importantly, these epigenetic features in iPSCs were not found in ESCs generated by nuclear transfer. Furthermore, recent genome-wide analysis using next-generation sequencing actually showed that iPSCs contained not only aberrant copy-number variations but also residual somatic DNA methylation patterns with a higher incidence than in pluripotent stem cells generated by nuclear transfer (Ma et al. 2014). Therefore, many problems remain before considering iPSCs as perfect pluripotent stem cells. Researchers who use iPSCs for disease modeling have to keep these iPSC imperfections in mind and confirm the validity of their experimental data using disease-specific iPSCs while exercising extreme caution.

1.7 Control Types of Disease Modeling with iPSCs

To analyze cells that are differentiated from patients' iPSCs, normal control cells are essential for comparison and for the accurate evaluation of the in vitro phenotype. It is therefore important to focus on the imperfections of experimental controls in modeling disease with iPSCs, whatever the method of generating iPSCs. As controls for disease modeling, ESCs, iPSCs derived from healthy donors and genome-edited isogenic iPSCs have been used in previous reports of disease modeling with human iPSCs (Inoue et al. 2014). However, the most appropriate type of control is unclear at present. Each type of control has advantages and disadvantages for disease modeling.

Existing human ESC lines and human iPSC lines are available by donation from specific research institutions, and using these cell lines can avoid the need for the additional generation of control cell lines. Even if iPSCs of healthy donors are newly generated as experimental controls, additional experimental technique is not required besides generating the iPSCs from patients. But, because even genomes of healthy donors may contain single nucleotide polymorphisms that may lead to a certain disease, ESC and iPSCs derived from healthy donors cannot serve as perfectly healthy controls. On the other hand, genome-edited isogenic controls make it possible to avoid the influence derived from cell line variations and single nucleotide polymorphisms in healthy donors. This requires reproducibility of results obtained from several control lines to avoid off-target effects. Furthermore, the additional culture required for cloning cell lines may lead to an increased risk of a genome abnormality occurring. Therefore, iPSC lines that serve as perfectly normal controls for disease modeling experiments are difficult to obtain at present.

1.8 Conclusion

Modeling of disease using patient-derived iPSCs is expected to drive medical advances, earlier than cell transplantation therapies using iPSCs that have the associated issue of recipient's safety to consider. Disease modeling with iPSCs at present has various uncertainties in the accuracy, reliability, and usability of methods. Researchers must advance their study of human disease while also recognizing these aspects associated with iPSC use. Analyzing human cells that have a disease phenotype in vitro is a groundbreaking approach and may possibly solve problems that have not yet been understood using animal experimental models. There is no doubt that modeling disease with iPSCs is a novel method for disease research and has the potential to cause considerable medical advancement.

References

Aasen T, Raya A, Barrero MJ, Garreta E, Consiglio A, Gonzalez F, Izpisua Belmonte JC (2008) Efficient and rapid generation of induced pluripotent stem cells from human keratinocytes. Nat Biotechnol 26(11):1276–1284. doi:10.1038/nbt.1503

Baker DE, Harrison NJ, Maltby E, Smith K, Moore HD, Shaw PJ, ... Andrews PW (2007) Adaptation to culture of human embryonic stem cells and oncogenesis in vivo. Nat Biotechnol 25(2):207–215. doi:10.1038/nbt1285

Choi SM, Liu H, Chaudhari P, Kim Y, Cheng L, Feng J, ... Jang YY (2011). Reprogramming of EBV-immortalized B-lymphocyte cell lines into induced pluripotent stem cells. Blood 118(7): 1801–1805. doi:10.1182/blood-2011-03-340620

Dang CV (2012) MYC on the path to cancer. Cell 149(1):22–35. doi:10.1016/j.cell.2012.03.003

Fusaki N, Ban H, Nishiyama A, Saeki K, Hasegawa M (2009) Efficient induction of transgene-free human pluripotent stem cells using a vector based on Sendai virus, an RNA virus that does not integrate into the host genome. Proc Jpn Acad B Phys Biol Sci 85(8):348–362

Garber K (2013) Inducing translation. Nat Biotechnol 31(6):483–486. doi:10.1038/nbt.2602

Gore A, Li Z, Fung HL, Young JE, Agarwal S, Antosiewicz-Bourget J, ... Zhang K (2011) Somatic coding mutations in human induced pluripotent stem cells. Nature 471(7336):63–67. doi:10.1038/nature09805

Green DR, Kroemer G (2009) Cytoplasmic functions of the tumour suppressor p53. Nature 458(7242):1127–1130. doi:10.1038/nature07986

Han J, Yuan P, Yang H, Zhang J, Soh BS, Li P, ... Lim B (2010) Tbx3 improves the germ-line competency of induced pluripotent stem cells. Nature 463(7284):1096–1100. doi:10.1038/nature08735

Hawley RG, Lieu FH, Fong AZ, Hawley TS (1994) Versatile retroviral vectors for potential use in gene therapy. Gene Ther 1(2):136–138

Hong H, Takahashi K, Ichisaka T, Aoi T, Kanagawa O, Nakagawa M, ... Yamanaka S (2009) Suppression of induced pluripotent stem cell generation by the p53-p21 pathway. Nature 460(7259):1132–1135. doi:10.1038/nature08235

Hussein SM, Batada NN, Vuoristo S, Ching RW, Autio R, Narva E, ... Otonkoski T (2011) Copy number variation and selection during reprogramming to pluripotency. Nature 471(7336): 58–62. doi:10.1038/nature09871

Inoue H, Nagata N, Kurokawa H, Yamanaka S (2014) iPS cells: a game changer for future medicine. EMBO J 33(5):409–417. doi:10.1002/embj.201387098

Kajiwara M, Aoi T, Okita K, Takahashi R, Inoue H, Takayama N, ... Yamanaka S (2012) Donor-dependent variations in hepatic differentiation from human-induced pluripotent stem cells. Proc Natl Acad Sci USA 109(31):12538–12543. doi:10.1073/pnas.1209979109

Kane NM, Nowrouzi A, Mukherjee S, Blundell MP, Greig JA, Lee WK, ... Baker AH (2010) Lentivirus-mediated reprogramming of somatic cells in the absence of transgenic transcription factors. Mol Ther 18(12):2139–2145. doi:10.1038/mt.2010.231

Kawamura T, Suzuki J, Wang YV, Menendez S, Morera LB, Raya A, ... Izpisua Belmonte JC (2009) Linking the p53 tumour suppressor pathway to somatic cell reprogramming. Nature 460(7259):1140–1144. doi:10.1038/nature08311

Kim D, Kim CH, Moon JI, Chung YG, Chang MY, Han BS, ... Kim KS (2009) Generation of human induced pluripotent stem cells by direct delivery of reprogramming proteins. Cell Stem Cell 4(6):472–476. doi:10.1016/j.stem.2009.05.005

Kim K, Doi A, Wen B, Ng K, Zhao R, Cahan P, ... Daley GQ (2010) Epigenetic memory in induced pluripotent stem cells. Nature 467(7313):285–290. doi:10.1038/nature09342

Lee AS, Tang C, Rao MS, Weissman IL, Wu JC (2013) Tumorigenicity as a clinical hurdle for pluripotent stem cell therapies. Nat Med 19(8):998–1004. doi:10.1038/nm.3267

Li H, Collado M, Villasante A, Strati K, Ortega S, Canamero M, ... Serrano M (2009) The Ink4/Arf locus is a barrier for iPS cell reprogramming. Nature 460(7259):1136–1139. doi:10.1038/nature08290

Loh YH, Hartung O, Li H, Guo C, Sahalie JM, Manos PD, ... Daley GQ (2010) Reprogramming of T cells from human peripheral blood. Cell Stem Cell 7(1):15–19. doi:10.1016/j.stem.2010.06.004

Ma H, Morey R, O'Neil RC, He Y, Daughtry B, Schultz MD, ... Mitalipov S (2014) Abnormalities in human pluripotent cells due to reprogramming mechanisms. Nature 511(7508):177–183. doi:10.1038/nature13551

Maekawa M, Yamaguchi K, Nakamura T, Shibukawa R, Kodanaka I, Ichisaka T, ... Yamanaka S (2011) Direct reprogramming of somatic cells is promoted by maternal transcription factor Glis1. Nature 474(7350):225–229. doi:10.1038/nature10106

Marion RM, Strati K, Li H, Murga M, Blanco R, Ortega S, ... Blasco MA (2009) A p53-mediated DNA damage response limits reprogramming to ensure iPS cell genomic integrity. Nature 460(7259):1149–1153. doi:10.1038/nature08287

Miura K, Okada Y, Aoi T, Okada A, Takahashi K, Okita K, ... Yamanaka S (2009) Variation in the safety of induced pluripotent stem cell lines. Nat Biotechnol 27(8):743–745. doi:10.1038/nbt.1554

Miyoshi K, Tsuji D, Kudoh K, Satomura K, Muto T, Itoh K, Noma T (2010) Generation of human induced pluripotent stem cells from oral mucosa. J Biosci Bioeng 110(3):345–350. doi:10.1016/j.jbiosc.2010.03.004

Nakagawa M, Koyanagi M, Tanabe K, Takahashi K, Ichisaka T, Aoi T, ... Yamanaka S (2008) Generation of induced pluripotent stem cells without Myc from mouse and human fibroblasts. Nat Biotechnol 26(1):101–106. doi:10.1038/nbt1374

Nakagawa M, Takizawa N, Narita M, Ichisaka T, Yamanaka S (2010) Promotion of direct reprogramming by transformation-deficient Myc. Proc Natl Acad Sci U S A 107(32):14152–14157. doi:10.1073/pnas.1009374107

Nishimura T, Kaneko S, Kawana-Tachikawa A, Tajima Y, Goto H, Zhu D, ... Nakauchi H (2013) Generation of rejuvenated antigen-specific T cells by reprogramming to pluripotency and redifferentiation. Cell Stem Cell 12(1):114–126. doi:10.1016/j.stem.2012.11.002

Okita K, Ichisaka T, Yamanaka S (2007) Generation of germline-competent induced pluripotent stem cells. Nature 448(7151):313–317. doi:10.1038/nature05934

Okita K, Matsumura Y, Sato Y, Okada A, Morizane A, Okamoto S, ... Yamanaka S (2011) A more efficient method to generate integration-free human iPS cells. Nat Methods 8(5):409–412. doi:10.1038/nmeth.1591

Okita K, Yamakawa T, Matsumura Y, Sato Y, Amano N, Watanabe A, ... Yamanaka S (2013) An efficient nonviral method to generate integration-free human-induced pluripotent stem cells from cord blood and peripheral blood cells. Stem Cells 31(3):458–466. doi:10.1002/stem.1293

Polo JM, Liu S, Figueroa ME, Kulalert W, Eminli S, Tan KY, ... Hochedlinger K (2010) Cell type of origin influences the molecular and functional properties of mouse induced pluripotent stem cells. Nat Biotechnol 28(8):848–855. doi:10.1038/nbt.1667

Seki T, Yuasa S, Oda M, Egashira T, Yae K, Kusumoto D, ... Fukuda K (2010) Generation of induced pluripotent stem cells from human terminally differentiated circulating T cells. Cell Stem Cell 7(1):11–14. doi:10.1016/j.stem.2010.06.003

Stadtfeld M, Nagaya M, Utikal J, Weir G, Hochedlinger K (2008) Induced pluripotent stem cells generated without viral integration. Science 322(5903):945–949. doi:10.1126/science.1162494

Staerk J, Dawlaty MM, Gao Q, Maetzel D, Hanna J, Sommer CA, ... Jaenisch R (2010) Reprogramming of human peripheral blood cells to induced pluripotent stem cells. Cell Stem Cell 7(1):20–24. doi:10.1016/j.stem.2010.06.002

Stewart CL, Stuhlmann H, Jahner D, Jaenisch R (1982) De novo methylation, expression, and infectivity of retroviral genomes introduced into embryonal carcinoma cells. Proc Natl Acad Sci U S A 79(13):4098–4102

Takahashi K, Yamanaka S (2006) Induction of pluripotent stem cells from mouse embryonic and adult fibroblast cultures by defined factors. Cell 126(4):663–676. doi:10.1016/j.cell.2006.07.024

Takahashi K, Tanabe K, Ohnuki M, Narita M, Ichisaka T, Tomoda K, Yamanaka S (2007) Induction of pluripotent stem cells from adult human fibroblasts by defined factors. Cell 131(5):861–872. doi:10.1016/j.cell.2007.11.019

Utikal J, Polo JM, Stadtfeld M, Maherali N, Kulalert W, Walsh RM, ... Hochedlinger K (2009) Immortalization eliminates a roadblock during cellular reprogramming into iPS cells. Nature 460(7259):1145–1148. doi:10.1038/nature08285

Vizcardo R, Masuda K, Yamada D, Ikawa T, Shimizu K, Fujii S, ... Kawamoto H (2013) Regeneration of human tumor antigen-specific T cells from iPSCs derived from mature CD8(+) T cells. Cell Stem Cell 12(1):31–36. doi:10.1016/j.stem.2012.12.006

Warren L, Manos PD, Ahfeldt T, Loh YH, Li H, Lau F, ... Rossi DJ (2010) Highly efficient reprogramming to pluripotency and directed differentiation of human cells with synthetic modified mRNA. Cell Stem Cell 7(5):618–630. doi:10.1016/j.stem.2010.08.012

Woltjen K, Michael IP, Mohseni P, Desai R, Mileikovsky M, Hamalainen R, ... Nagy A (2009) PiggyBac transposition reprograms fibroblasts to induced pluripotent stem cells. Nature 458(7239): 766–770. doi:10.1038/nature07863

Yamanaka S (2010) Patient-specific pluripotent stem cells become even more accessible. Cell Stem Cell 7(1):1–2. doi:10.1016/j.stem.2010.06.009

Zhou H, Wu S, Joo JY, Zhu S, Han DW, Lin T, ... Ding S (2009) Generation of induced pluripotent stem cells using recombinant proteins. Cell Stem Cell 4(5):381–384. doi:10.1016/j.stem.2009.04.005

Cardiomyopathy

2

Studying Broken Hearts in a Dish...

Elena Matsa, Karim Sallam, and Joseph C. Wu

Abstract

Cardiomyopathies are mechanical disorders of the heart muscle that can lead to heart failure and lethal arrhythmias. Still accounting for over 20,000 deaths annually in the United States alone, these disorders are often misdiagnosed and mistreated. The ability to generate human heart cells in vitro via the use of induced pluripotency technologies is now allowing researchers to study cardiomyopathies caused by known genetic mutations. This chapter describes the clinical features, classifications and current treatments for cardiomyopathy and provides a guide of how cardiac cells generated from patient-specific human induced pluripotent stem cells (hiPSCs) have been used to elucidate important pathological mechanisms and propose novel treatments for this class of lethal cardiac conditions.

Keywords

Cardiomyopathy • Cardiomyocyte • Induced pluripotency • Stem cells • Disease modelling

E. Matsa, Ph.D. • J.C. Wu, M.D., Ph.D. (✉)
Stanford Cardiovascular Institute, Stanford University School of Medicine, Stanford, CA, USA
e-mail: joewu@Stanford.edu

K. Sallam, M.D.
Department of Medicine, Division of Cardiology, Stanford University School of Medicine, Lorry I. Lokey Stem Cell Research Building, 265 Campus Drive, Rm G1120, Stanford, CA 94305-5111, USA

K. Fukuda (ed.), *Human iPS Cells in Disease Modelling*,
DOI 10.1007/978-4-431-55966-5_2

2.1 Introduction to Clinical Features of Cardiomyopathy

Cardiomyopathy is a pathologic disorder of the heart muscle that affects millions of individuals and accounts for approximately 23,000 deaths annually in the United States (NHLBI 2012). It is characterised by defects in the mechanical function of the heart, which often lead to heart failure syndrome or sudden cardiac death due to increased predisposition to lethal arrhythmias. Several classes of cardiomyopathy have been defined, mainly distinguished by their pathologic findings and clinical course, which can be attributed to inherited or acquired causes. It is well documented that acquired cardiomyopathy leads to myocyte injury via ischemia, increased loading conditions or systemic disorders affecting the heart, but the pathophysiology of inherited cardiomyopathy remains poorly understood. Current knowledge of disease classifications, etiologies, diagnosis, risk stratification and treatment of cardiomyopathy are further detailed in this introductory section (Table 2.1).

Table 2.1 List of abbreviations

Abbreviation	Explanation
ACTN2	Alpha 2-actinin
ANF	Atrial natriuretic factor
ARVC	Arrhythmogenic right ventricular cardiomyopathy
CALN	Calcineurin
CASP3	Caspase 3
CM	Cardiomyocyte
DAD	Delayed afterdepolarisation
DCM	Dilated cardiomyopathy
EB	Embryoid body
GWAS	Genome-wide association studies
HCM	Hypertrophic cardiomyopathy
hESC	Human embryonic stem cell
hiPSC	Human induced pluripotent stem cell
hPSC	Human pluripotent stem cell
ICD	Implantable cardioverter defibrillators
MEA	Multielectrode array
MYH	Myosin heavy chain
NFAT	Nuclear factor of activated T cells
PKP2	Plakophilin-2
PPAR	Peroxisome proliferator-activated receptor
RCM	Restrictive cardiomyopathy
SERCA2a	Sarcoplasmic reticulum Ca2+ adenosine triphosphatase
TNNC	Troponin C
TNNT2	Troponin T type 2

2.1.1 Disease Classification

Multiple classification systems have been proposed for cardiomyopathy, but the most widely adopted system is based on structural and pathologic features of the disorder that tend to also dictate much of the therapeutic approach for each class (Richardson et al. 1996). As illustrated in Fig. 2.1, *hypertrophic cardiomyopathy* (HCM) is notable for increased relative thickness of (mostly) the left ventricle, which is often asymmetric and associated with hemodynamic significance. *Dilated cardiomyopathy* (DCM) is characterised by progressive dilation of one or both ventricular cavities, whereas *restrictive cardiomyopathy* (RCM) is characterised by impaired diastolic filling of one or both ventricles. *Arrhythmogenic right ventricular cardiomyopathy* (ARVC) is recognised most commonly by fibrofatty infiltration of the right ventricle leading to arrhythmic disorders and sudden death.

The same classification system also identifies certain subgroups as **specific cardiomyopathies**, i.e. cardiomyopathies in which a clear underlying cause is defined but is often unrelated to cardiac muscle. Some of these disorders include ischemic, hypertensive and toxic cardiomyopathy, whose pathologies and treatments may overlap with DCM, HCM or RCM while still maintaining features specific to the disorder.

Some cardiomyopathies such as *noncompaction cardiomyopathy* are *unclassified* under the various categories described above. To fill this gap, some researchers have proposed classifying cardiomyopathies as being *primary*, *secondary* or *mixed*, based on underlying molecular and cellular pathophysiology causing the disorder rather than on end-stage features. Under this system, primary cardiomyopathies are those involving the heart and can be genetic, acquired or both, whereas secondary cardiomyopathies involve the heart as part of a broader systemic disorder. This new classification system may enable identification of common pathways involved in various cardiomyopathies, potentially improving our understanding of the disorders and leading to better therapies (Maron et al. 2006).

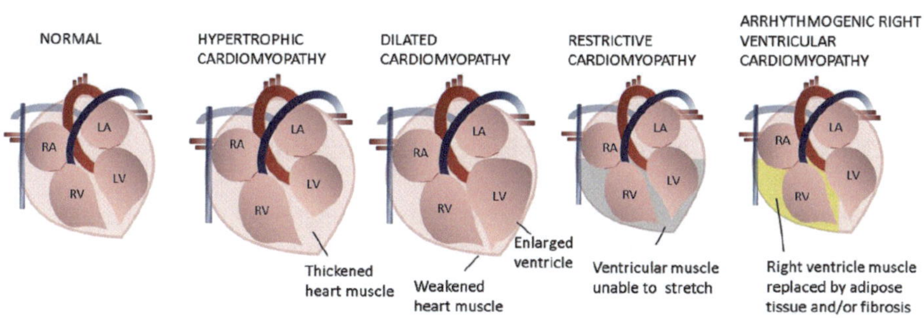

Fig. 2.1 Classification of cardiomyopathies. Schematic diagram depicting structural changes that occur in cardiac muscle during cardiomyopathy. In *dilated cardiomyopathy*, progressive weakness in ventricular muscle is observed, accompanied by enlargement of one or both ventricular cavities, whereas in *hypertrophic cardiomyopathy*, enlargement is mostly in the left ventricular muscle, constricting the ventricle(s). *Restrictive cardiomyopathy* is characterised by impaired diastolic filling of one or both ventricles. Finally, *arrhythmogenic right ventricular cardiomyopathy* is typically characterised by fibrofatty deposition in the right ventricle

2.1.2 Familial Cardiomyopathy

Familial cardiomyopathies are those of genetic origin, frequently identified as "primary" in the most recent classification system (Sect. 2.1.1). Over 2,000 mutations in 30 genes have thus far been implicated in cardiomyopathy, including the genes troponin C, T and I (*TNNC1*, *TNNT2*, *TNNI3*), myosin-6 and myosin-7 (*MYH6*, *MYH7*), α-actinin (*ACTN2*), lamin (*LMNA/C*) and dystrophin (*DMD*) (Maron and Maron 2013). Although inheritance is typically autosomal dominant, X-linked and autosomal recessive patterns have also been observed (Hershberger et al. 2010; Maron et al. 2012). HCM is most commonly linked to sarcomeric gene mutations which together account for 75 % of identifiable inherited HCMs. Familial cases of RCM are mainly attributed to *cTNNI3* gene mutations (Sen-Chowdhry et al. 2010), whereas DCM is far more genetically heterogeneous, with rarer mutations also found in cytoskeletal, nucleoskeletal, mitochondrial and calcium-handling protein-encoding genes (McNally et al. 2013). Over 50 % of ARVC cases are linked to mutations in desmosome protein-encoding genes (e.g. plakophilin-2, *PKP2*; plakoglobin, *PKG*; desmoplakin; *DSP*) (Moric-Janiszewska and Markiewicz-Loskot 2007).

Because relatively few clinical clues guide the diagnosis of inherited cardiomyopathies, genetic testing can help identify patients at risk of accelerated disease progression, congestive heart failure and arrhythmia (Ackerman et al. 2011). However, over 70 % of patients with familial DCM and 40 % of patients with HCM have no identifiable genetic testing variants (Hershberger et al. 2010; Lakdawala et al. 2012; Van Driest et al. 2005). Such cardiomyopathy cases are considered *idiopathic*, a decision that potentially bears huge prognostic and diagnostic implications for affected individuals and family members. With recent arrival of next-generation sequencing methods and completion of the Human Genome Project, genome-wide association studies (GWAS) have enabled the identification of a growing number of genetic variants as myopathy-causing. This has changed the cases of DCM thought to be familial from 2 % in 1982 to 25–30 % in 2012 (Lenfant 2013), allowing timely screening of family members and early initiation of therapy (Hershberger and Siegfried 2011).

2.1.3 Clinical Diagnosis and Risk Stratification

The diagnosis of cardiomyopathy is based on a constellation of clinical findings and tests, including pathological examinations, electrocardiography and echocardiography. Advanced imaging modalities such as magnetic resonance imaging (MRI) have enhanced our ability to identify pathological changes in the absence of symptoms or when the clinical presentation is unclear (Steeds 2013). Once diagnosis is made, patients can have variable clinical courses in terms of response to therapy and risk for arrhythmic complications. A major challenge in treating all classes of cardiomyopathy is the ability to stratify the arrhythmic risk associated with each given case. General clinical and imaging parameters guide the use of implantable cardioverter defibrillators (ICDs) in DCM and HCM (Goldberger et al. 2011; Epstein et al. 2013), but while some of these guidelines work well on a population level, data show that only 17–64 % of patients receive appropriate shocks (Germano et al. 2006). Notably, at least 30 % of patients suffer risks associated with ICD implantation without benefit. This creates a significant cost burden for the health care system, and highlights the clear need for improved risk stratification tools to better identify arrhythmic risks in cardiomyopathy patients.

2.1.4 Currently Available Therapies

The mainstay of medical therapy for DCM includes β-blockade, angiotensin-converting enzyme inhibition or angiotensin receptor blockade and aldosterone antagonists. Inhibition of catecholamine and renin-angiotensin-aldosterone axis effects on the heart, combined with mechanical unloading, has also provided substantial improvement in patient survival on a population level (Jessup et al. 2009). Nevertheless, current treatment mainly relies on retarding deleterious cardiac remodelling and providing symptomatic relief through modulating the hemodynamic consequences of myopathy such as reducing preload or reducing intracardiac obstruction, rather than combating disease causation (Gersh et al. 2011; Kushwaha et al. 1997). Moreover, there is a vast range of responsiveness to pharmacotherapy, with some patients experiencing near complete recovery and others deriving little or no benefit.

For other classes of cardiomyopathy such as HCM and RCM, there has been little progress in pharmacotherapy that offers significant mortality improvement. Therefore, substantial efforts are needed to identify more effective drug treatments for these types of cardiomyopathy. For example, it would be highly beneficial to rapidly identify patients who might not respond to traditional therapy, so they could be timely referred to cardiac transplant or ICD therapy (Parry et al. 2013). This would also reduce the $40 billion socioeconomic burden currently spent in management of cardiomyopathies (Heidenreich et al. 2011).

2.2 Animal Models of Cardiomyopathy

As mentioned earlier, the precise disease mechanisms of familial cardiomyopathy are not well comprehended. This poses major challenges in understanding cellular and systemic pathophysiologies involved in cardiomyopathy and identifying potential targets for treatment. Laboratory investigations of cardiomyopathy could shed light into these mechanisms, but have been limited by the inability to obtain and sustainably culture adult human heart cells. Genetically modified mice are the most widely utilised research animals for studying cardiomyopathy, and have thus far allowed mechanistic studies on DCM, HCM and ARVC (Recchia and Lionetti 2007; Maass and Leinwand 2000; Pilichou et al. 2011). These models have proven to be critical in the discovery of novel pathophysiological mechanisms, but have offered little advancement in diagnosis and therapy. This is attributed to differences in gene expression and physiology between humans and animal species. For example, the murine heart beats nearly eight times faster than the human (500 bpm vs. 60 bpm) (Doevendans et al. 1998). Furthermore, mouse and human hearts rely on separate sets of ion channels for their contraction and relaxation and exhibit differences in use of regulatory proteins (e.g. phospholamban) (Davis et al. 2011), distribution of structural genes (e.g. alpha- and beta-myosin heavy chains, α-/β-MHC) (Morano 1999) and expression of surface markers (e.g. SIRPA) (Dubois et al. 2011). Such differences mean that extrapolation of research findings from mouse to human can be tenuous, something that has long posed need for development of novel in vitro humanised systems for studying cardiomyopathies.

2.3 De Novo Generation of Human Cardiomyocytes Using Pluripotent Stem Cells

Recent discovery of hiPSC technologies has now brought us closer to human laboratory models of cardiomyopathy. Tissue samples (e.g. skin, blood) can be collected from cardiomyopathy patients with relative ease and subsequently converted to human cardiomyocytes retaining appropriate genetic variant(s) to closely mimic disease pathology. This ability may offer insights into cardiomyopathy that could identify novel diagnostic, pharmacogenomic and therapeutic targets. Methods for generating and studying hiPSC-derived cardiomyocytes (hiPSC-CMs) in the laboratory are further discussed in this section.

2.3.1 Methods for In Vitro Cardiac Differentiation

Similar to pluripotent human embryonic stem cells (hESCs), hiPSCs have the remarkable potential of differentiation into many cell types of the adult body (Reubinoff et al. 2000; Takahashi et al. 2007), including spontaneously contracting cardiomyocytes (Burridge et al. 2007). During in vitro differentiation, the first crucial step towards cardiogenesis is mesoderm formation, a step heavily dependent on NODAL and BMP4 signalling pathways (Mordwinkin et al. 2013). Cardiac specification then proceeds with formation of committed cardiac mesoderm and subsequently cardiac progenitors of the first and second heart fields (giving rise to endothelium and myocardium, and cardiac chambers, respectively) (Buckingham et al. 2005). Finally, progenitors mature into cardiomyocytes that can be identified by spontaneous contraction and expression of sarcomeric proteins, such as TNNT2 and ACTN2 (Rajala et al. 2011). Three key methods are available for mesoderm induction and cardiac specification of pluripotent stem cells to cardiomyocytes (Burridge et al. 2012). These include (1) coculture with END2 mouse visceral endoderm-like cells, a relatively inefficient method that nevertheless provides ~85 % ventricular cardiomyocytes (Mummery et al. 2003); (2) embryoid body (EB) formation via forced aggregation in 96-well plates, an efficient but technically demanding method; and (3) monolayer differentiation of pluripotent cultures, which is to date the most efficient, reproducible and scalable method available (Burridge et al. 2012). Growth factors Activin-A, BMP4 and FGF2 have been shown to enhance cardiac specification of hiPSCs, in a time and dose-dependent manner (Kattman et al. 2011; Xu et al. 2011). It has also been shown that growth factors can be replaced by small molecules such as CHIR99021, a selective inhibitor of glycogen synthase kinase 3 beta (GSK3B) that activates the canonical WNT signalling pathway, thus inducing mesoderm formation (Hu et al. 2013).

Cardiac specification of pluripotent stem cells was until recently a largely inefficient process, generating low purity populations of cardiomyocytes that need to be further refined by either labour-intensive genetic selection processes or use of surface markers (e.g. SIRPA) and mitochondrial dyes (e.g. MitoTracker) (Matsa and Denning 2012). More recent protocols relying on monolayer differentiation into cardiomyocytes have alleviated the need for purification as they typically yield >90 % cardiomyocytes. This has been a significant step towards generating high quality cardiomyocyte populations that can be used in disease modelling and drug screening. Generating purified populations of each cardiomyocyte subtype, however, presents a major remaining hurdle as in vitro preparations of cardiac myocytes typically contain proportions of pacemaker and working chamber (atrial and ventricular) cells.

2.3.2 In Vitro Characterisation of Cardiomyocytes

The use of in vitro-derived hiPSC-CMs for disease modelling requires their characteristics to be physiologically analogous to human cardiomyocytes in vivo. Gene and protein expression assays are typically performed to assure contracting cells form striated sarcomeres and express terminal cardiac markers and structural proteins such as ACTN2, TNNT2, MYH6 and MYH7. In vitro characterisation also relies on detailed electrophysiological analysis, mainly comprising patch-clamp and multielectrode arrays (MEAs; see Box 2.1) (Navarrete et al. 2013; Matsa et al. 2011). These methods can be exploited to determine the percentage of pacemaker-, atrial- and ventricular-like cells that are formed, as well as to measure drug effects on contractility and ion channel currents. Calcium transients (see Box 2.1), an important property for contraction of cardiomyocytes, can also be visualised in the laboratory using real-time microscopy in the presence of calcium-sensitive dyes (Jung et al. 2012).

Box 2.1: Electrophysiology Terms

Multielectrode arrays (MEAs) are laboratory electronic devices containing a set of microelectrode wires in a fixed spatial arrangement, that are capable of detecting voltage changes in extracellular environment/culture medium caused by excitation of cells adhering to the surface of the devices.

(continued)

Box 2.1 (continued)

Patch-clamp is a technique used in laboratory electrophysiology in which a single excitable cell is perforated with a glass needle to measure changes in membrane voltage, in order to study the electrical behaviour of single ion channels found across cell membranes.

Excitation-contraction coupling is the relationship between the electrical depolarisation of the cardiac cell membrane and the activation of the contractile myofilaments.

Calcium transients refer to the amount of calcium ions (Ca^{2+}) present for a short time period in the sarcoplasm of an excitable cell, while the cell undergoes cycles of excitation and relaxation.

Chronotropic effect (Greek chronos = time) is one that changes the rate of muscle contraction.

Inotropic effect (Greek ina = fibre) is one that changes the force of muscle contraction.

Previous molecular and electrophysiological studies have revealed that hiPSC-CMs have normal cardiomyocyte functional properties, including contraction regulated by physiologic intracellular signalling such as excitation-contraction coupling (Zhang et al. 2009; Anderson et al. 2007). However, cells are also reported to have some immature developmental properties; for instance, their spontaneous contraction in culture is an indication of immature automaticity typically observed in fetal hearts. They also exhibit fetal-type ion channel expression, electrophysiological signals, gene expression patterns and physical phenotypes (Davis et al. 2011; Cao et al. 2008). Prolonged passaging of 3 months up to 1 year and temporal 3D culture have been shown to moderately increase maturation (Fu et al. 2008; Otsuji et al. 2010) by improving myofibril density and alignment and increasing the fraction of multinucleated cardiomyocytes observed (Lundy et al. 2013). However, more cost and time-effective methods for maturation have not been described to date. Regardless of their immaturity, the potential ability of hiPSC-CMs to "capture the entire patient's genome in a dish" offers unique opportunities to identify genetic loci or cellular pathways related to predisposition towards cardiac disorders.

2.4 Recapitulating Cardiomyopathies in a Dish

Just 8 years after the first report of somatic cell reprogramming to pluripotency, hiPSCs have been used to generate over 80 models of human disease. Conditions affecting the heart, smooth muscle, skeletal muscle, immune system, skin, central nervous system, blood and eye, as well as imprinting, metabolic and multi-organ disorders have been recapitulated in the lab and used to gain insightful information regarding disease mechanism and potential novel therapies (Rajamohan et al. 2013; Matsa et al. 2014). This section focuses on reports of cardiomyopathy disease modelling using cardiomyocytes derived from hiPSCs as an in vitro platform.

2.4.1 Dilated Cardiomyopathy

In 2012, Sun et al. were able to obtain skin tissue samples from DCM family members carrying a p.Arg173Trp point mutation in the gene-encoding sarcomeric cardiac TNNT2 (Sun et al. 2012). Following the reprogramming of skin cells to hiPSCs, they generated cardiomyocytes which, compared to control genetically matched healthy samples, exhibited irregular organisation of sarcomeric cardiac ACTN2, scattered distribution pattern of Z bodies, reduction in contractile force and altered regulation of calcium ions (Ca^{2+}), as observed by calcium imaging. DCM hiPSC-CMs also expressed reduced levels of calcium related key molecules, such as CASQ, TMEM38, NFAT and NECAB. When challenged with the β-adrenergic agonist, norepinephrine, DCM hiPSC-CMs showed increased susceptibility to cellular stress. Using patch-clamp electrophysiology, susceptibility was reported as a positive inotropic and negative chronotropic effect (see Box 2.1). Fluorescence imaging determined that adrenergic stimulation also increased the number of cells with abnormal sarcomeric ACTN2 distribution. These findings demonstrated that hiPSC-CMs from DCM patients could closely recapitulate the morphological and functional phenotypes of DCM.

Furthermore, the same study demonstrated that prolonged treatment with the β-adrenergic blocker, metoprolol, or overexpression of sarcoplasmic reticulum Ca^{2+} adenosine triphosphatase (SERCA2a) improved the function of DCM hiPSC-CMs, causing negative chronotropic effects and improving global Ca^{2+} transients. It was suggested that metoprolol might lead to reduction in contraction force as well as contraction frequency, thus alleviating cardiomyocytes of mechanical stress. Microarray analysis demonstrated that gene therapy by SERCA2a overexpression was capable of improving abnormal cardiomyocyte function by acting on the Ca^{2+}, protein kinase K, G-protein coupled receptor, integrin, cytoskeletal and ubiquitination signalling pathways. Although it is still not clear whether altered Ca^{2+} handling is the primary factor that contributes to the disease or merely a secondary consequence of disease progression, the study demonstrated that hiPSC-CMs can provide an important platform to investigate treatments that might clinically benefit DCM disease cardiomyocyte function in culture and further refine our understanding of specific disease mechanisms of DCM.

2.4.2 Hypertrophic Cardiomyopathy

Similar to DCM discussed above, Lan et al. were able to recapitulate the pathophysiology of HCM using induced pluripotency technologies. They obtained skin samples from a ten-member family cohort carrying a hereditary HCM-related missense mutation (p.Arg663His) in MYH7. Diseased cardiomyocytes generated from hiPSCs were able to mimic numerous aspects of the HCM clinical characteristics, including 60 % cell enlargement and 26 % increased multinucleation, 105 % increased myofibril content, increased expression of atrial natriuretic factor (ANF), elevation of beta-myosin (MYH7)/alpha-myosin

(MYH6) ratio, calcineurin (CALN) activation and nuclear localisation of nuclear factor of activated T cells (NFAT), as evidenced by immunofluorescence staining. Patch-clamp electrophysiology revealed contractile arrhythmias in the form of delayed afterdepolarisations (DADs) at the single-cell level, whereas calcium imaging demonstrated perturbations in Ca^{2+} cycling and elevation in intracellular Ca^{2+} concentrations. To confirm the p.Arg663His mutation caused the disease pathology, the study showed that normal hESC-CMs genetically engineered to carry the same mutation were able to recapitulate the calcium-handling abnormalities of HCM hiPSC-CMs.

Prolonged treatment of HCM hiPSC-CMs with the β-adrenergic agonist, isoproterenol, was found to provoke Ca^{2+} transient irregularities and arrhythmias, mirroring the development of arrhythmias in HCM patients under sympathetic stimulation. Importantly, co-administration of the β-adrenergic blocker, propranolol, with isoproterenol significantly reduced catecholamine-induced exacerbation of hypertrophy, Ca^{2+} handling deficiencies and arrhythmia. Moving one step further, Lan et al. were able to show that pharmacological restoration of Ca^{2+} homeostasis with the L-type Ca^{2+} channel blocker, verapamil, prevented development of cell enlargement and electrophysiological abnormalities completely. The investigators also tested 13 pharmacological agents on HCM iPSC-CMs. Only those capable of blocking Ca^{2+} and Na^{+} entry alleviated DADs in HCM hiPSC-CMs. It is believed that reduction of Na^{+} influx limits Ca^{2+} concentration by allowing Na^{+}/Ca^{2+} exchange to remove Ca^{2+} more readily. Therefore, these results showed that perturbations in Ca^{2+} cycling and elevation in intracellular Ca^{2+} concentrations can be considered as central mechanisms for disease development at the cellular level, and demonstrated the potential of patient-specific hiPSC-CMs as a powerful tool for identification of novel pharmaceutical agents to treat HCM.

A separate study was able to show that hiPSC-CMs generated from patients suffering with *LEOPARD syndrome*, 80 % of whom present hypertrophic cardiomyopathy as the most life-threatening aspect of the disorder, had a significantly increased median surface, higher degree of sarcomeric organisation, and 50 % increased preferential localisation of nuclear factor of activated T cells (NFATC4) in the nucleus when compared with hESC-CMs or hiPSC-CMs from healthy family members of the patients (Carvajal-Vergara et al. 2010). These features correlate well with the potential hypertrophic state observed in LEOPARD syndrome patients. Using a phosphoproteomic microarray chip containing approximately 600 pan and phospho-specific antibodies, this study also demonstrated that phosphorylation perturbations in proteins involved in the RAS-MAPK signal transduction pathway might be involved in development of disease pathophysiology. Once again, these findings demonstrate that hiPSC-CMs can provide the required characteristics to precisely determine the pathology behind these familial hypertrophic disorders, laying a foundation for studying the molecular mechanisms of these diseases and investigating novel treatment interventions.

2.4.3 Arrhythmogenic Right Ventricular Cardiomyopathy

Modelling of ARVC using induced pluripotency technologies has been somewhat more challenging than for DCM and HCM above. This is because, in many cases, ARVC is an adult-onset disorder with a median age of presentation of 26 years that would typically not be manifested in the fetal-like cardiomyocytes generated from hiPSCs. Indeed, hiPSC-CMs derived from ARVC patients carrying plakophilin-2 (*PKP2*) mutations c.C2484T and c.2013delC were only able to recapitulate mild disease pathologies in vitro, such as abnormal plakoglobin nuclear translocation and decreased β-catenin activity. Induction of adult-like metabolic energetics in hiPSC-CMs from an embryonic glycolytic state to an adult fatty acid oxidation state, via the use of an adipogenic cell culture environment, produced a 31.6 % increase in lipogenesis and 39.6 % increase in apoptosis of ARVC hiPSC-CMs compared to unaffected controls. In conjunction, expression of genes involved in fatty acid oxidation, such as peroxisome proliferator-activated receptor-α and -γ (*PPAR-α* and *-γ*) was increased, whereas pro-survival genes such as Caspase 3 (*CASP3*) were detected at lower levels. Thus, this model sufficiently recapitulated the exaggerated lipogenesis and apoptosis features observed clinically in ARVC patients. Furthermore, treatment of hiPSC-CMs with *PPAR-α* and *-γ* antagonists, GW9662, T0070907 and GW6471, was able to prevent lipogenesis and apoptosis in ARVC hiPSC-CMs, largely reverting ARVC pathogenesis features. The study concluded that induction of adult-like metabolic energetics in hiPSC-CMs, together with abnormal *PPAR-α* and *-γ* activation, underlies the pathogenesis of ARVC. Significantly, this was the first report to demonstrate modelling of an adult-onset disease using patient-specific hiPSCs and reveal crucial mechanistic insights of disease pathologies that enabled the proposal of novel disease-modifying therapeutic strategies.

2.5 Conclusions and Future Directions

This chapter outlines clinical manifestations of cardiomyopathies, as well as difficulties currently facing physicians when classifying and treating these disorders. It also highlights the three major classes of familial cardiomyopathy that thus far have been successfully recapitulated in vitro, with observed phenotypes closely resembling clinical symptoms described in patients. These studies show that hiPSC-based models of cardiomyopathy provide suitable tools for investigation of pathophysiological mechanisms, as well as for identification of novel drug therapies that could ameliorate observed pathologies and benefit future management and risk stratification for cardiomyopathies.

Another important use for hiPSC-based models of cardiomyopathy has been in drug toxicity screening. Notably, HCM iPSC-CMs (Liang et al. 2013) have been used to demonstrate increased susceptibility to arrhythmias when treated with the drug cisapride, now withdrawn from the market due to adverse side effects in cardiac repolarisation. Similarly, DCM hiPSC-CMs were found to exhibit drug-induced proarrhythmias when treated with high doses of Nicorandil, a drug clinically used for treatment of angina that presents palpitations as adverse side effects. Thus, disease-specific hiPSC-CMs demonstrate higher susceptibilities to cardiotoxicity as compared to healthy hiPSC-CMs. This correlates well with observations that patients with HCM and DCM are particularly sensitive to drug-induced cardiotoxicity, and are susceptible to lethal arrhythmias. Therefore, disease-specific hiPSC-CMs are not only a good model for disease modelling and drug discovery but also can serve as a valuable platform for drug toxicology studies that could provide timely and accurate prediction of adverse drug side effects on the cardiovascular system. This should benefit the pharmaceutical industry by reducing drug withdrawal at preclinical and clinical stages, as well as cardiomyopathy patients by increasing confidence in safety and efficacy of administered drugs (Rajamohan et al. 2013).

Major efforts are under way to identify more sophisticated subtype differentiation, maturation and purification protocols for hiPSC-CMs that can create accurate physiologic cellular conditions to reflect human adult disease phenotypes. Steady progress is being made towards this direction, and it is anticipated that future studies will shed light on details regarding the precise cellular pathways which are impaired in cardiomyopathy, making possible the identification of novel therapeutic small molecules, including large scale screening of chemical libraries. Developing scaled automated platforms for hiPSC-CM generation could enable future assessment of drug toxicity at an industrial level and bring us closer to generation of cardiac myocytes in appropriate quantity and quality for autologous (hiPSC-CM) and allogeneic (hESC-CM) clinical transplantations.

Finally, recent advances in gene-editing technologies utilising nuclease enzymes to enable targeted DNA modification for gene disruptions or genetic repair are also anticipated to significantly alter the field of disease modelling (Wang et al. 2012; Soldner and Jaenisch 2012). Advantages of these technologies include the ability to correct or introduce disease-causing mutations in hiPSCs to allow creation of genetically matched experimental controls. This could substantially simplify analysis of correlation between genomic variants and disease phenotypes. Along with the dawn of next-generation sequencing technologies, it is anticipated that these tools will enable identification of novel mutations implicated in cardiomyopathy and help explain why mutations in distinct locations of the same gene (e.g. *TNNT2*) might preferentially cause different types of cardiomyopathy. Such studies hold the potential to substantially improve future phenotype-to-genotype correlation analysis and clinical treatments for cardiomyopathies, as well as for other cardiovascular disorders.

Acknowledgements We gratefully acknowledge funding support from the PCBC Jump Start Award PCBC_JS_2013/3_02 and the AHA Beginning Grant in Aid 16BGIA27790017 (EM), the Burroughs Wellcome Foundation, the Leducq Foundation, NIH R01 HL123968, NIH R01 HL130020, NIH R01 HL126527, and the California Institute for Regenerative Medicine (CIRM) RT3-07798 and DR2-05394 (JCW).

References

Ackerman MJ, Priori SG, Willems S, Berul C, Brugada R, Calkins H, Camm AJ, Ellinor PT, Gollob M, Hamilton R, Hershberger RE, Judge DP, Le Marec H, McKenna WJ, Schulze-Bahr E, Semsarian C, Towbin JA, Watkins H, Wilde A, Wolpert C, Zipes DP (2011) HRS/EHRA expert consensus statement on the state of genetic testing for the channelopathies and cardiomyopathies: this document was developed as a partnership between the Heart Rhythm Society (HRS) and the European Heart Rhythm Association (EHRA). Europace 13(8):1077–1109

Anderson D, Self T, Mellor IR, Goh G, Hill SJ, Denning C (2007) Transgenic enrichment of cardiomyocytes from human embryonic stem cells. Mol Ther 15(11):2027–2036

Buckingham M, Meilhac S, Zaffran S (2005) Building the mammalian heart from two sources of myocardial cells. Nat Rev Genet 6(11):826–837

Burridge PW, Anderson D, Priddle H, Barbadillo Munoz MD, Chamberlain S, Allegrucci C, Young LE, Denning C (2007) Improved human embryonic stem cell embryoid body homogeneity and cardiomyocyte differentiation from a novel V-96 plate aggregation system highlights interline variability. Stem Cells 25(4):929–938

Burridge PW, Keller G, Gold JD, Wu JC (2012) Production of de novo cardiomyocytes: human pluripotent stem cell differentiation and direct reprogramming. Cell Stem Cell 10(1):16–28

Cao F, Wagner RA, Wilson KD, Xie X, Fu JD, Drukker M, Lee A, Li RA, Gambhir SS, Weissman IL, Robbins RC, Wu JC (2008) Transcriptional and functional profiling of human embryonic stem cell-derived cardiomyocytes. PLoS One 3(10):e3474

Carvajal-Vergara X, Sevilla A, D'Souza SL, Ang YS, Schaniel C, Lee DF, Yang L, Kaplan AD, Adler ED, Rozov R, Ge Y, Cohen N, Edelmann LJ, Chang B, Waghray A, Su J, Pardo S, Lichtenbelt KD, Tartaglia M, Gelb BD, Lemischka IR (2010) Patient-specific induced pluripotent stem-cell-derived models of LEOPARD syndrome. Nature 465(7299):808–812

Davis RP, van den Berg CW, Casini S, Braam SR, Mummery CL (2011) Pluripotent stem cell models of cardiac disease and their implication for drug discovery and development. Trends Mol Med 17(9):475–484

Doevendans PA, Daemen MJ, de Muinck ED, Smits JF (1998) Cardiovascular phenotyping in mice. Cardiovasc Res 39(1):34–49

Dubois NC, Craft AM, Sharma P, Elliott DA, Stanley EG, Elefanty AG, Gramolini A, Keller G (2011) SIRPA is a specific cell-surface marker for isolating cardiomyocytes derived from human pluripotent stem cells. Nat Biotechnol 29(11):1011–1018

Epstein AE, DiMarco JP, Ellenbogen KA, Estes NA 3rd, Freedman RA, Gettes LS, Gillinov AM, Gregoratos G, Hammill SC, Hayes DL, Hlatky MA, Newby LK, Page RL, Schoenfeld MH, Silka MJ, Stevenson LW, Sweeney MO, Tracy CM, Darbar D, Dunbar SB, Ferguson TB Jr, Karasik PE, Link MS, Marine JE, Shanker AJ, Stevenson WG, Varosy PD (2013) 2012 ACCF/AHA/HRS focused update incorporated into the ACCF/AHA/HRS 2008 guidelines for device-based therapy of cardiac rhythm abnormalities: a report of the American College of Cardiology Foundation/American Heart Association Task Force on Practice Guidelines and the Heart Rhythm Society. J Am Coll Cardiol 61(3):e6–e75

Fu J-D, Tse H-f, Siu C-W, Moore JC, Lieu DK, Liao S-Y, Lai W-H, Timofeyez V, Li N, Chiamvimonvat N, Li RA (2008) Driven maturation of embryonic stem cell-derived cardiomyocytes confers post-transplantation safety. Cell Res 18(S1):S132–S132

Germano JJ, Reynolds M, Essebag V, Josephson ME (2006) Frequency and causes of implantable cardioverter-defibrillator therapies: is device therapy proarrhythmic? Am J Cardiol 97(8): 1255–1261

Gersh BJ, Maron BJ, Bonow RO, Dearani JA, Fifer MA, Link MS, Naidu SS, Nishimura RA, Ommen SR, Rakowski H, Seidman CE, Towbin JA, Udelson JE, Yancy CW (2011) 2011 ACCF/AHA guideline for the diagnosis and treatment of hypertrophic cardiomyopathy: a report of the American College of Cardiology Foundation/American Heart Association Task Force on Practice Guidelines. Circulation 124(24):e783–e831

Goldberger JJ, Buxton AE, Cain M, Costantini O, Exner DV, Knight BP, Lloyd-Jones D, Kadish AH, Lee B, Moss A, Myerburg R, Olgin J, Passman R, Rosenbaum D, Stevenson W, Zareba W, Zipes DP (2011) Risk stratification for arrhythmic sudden cardiac death: identifying the roadblocks. Circulation 123(21):2423–2430

Heidenreich PA, Trogdon JG, Khavjou OA, Butler J, Dracup K, Ezekowitz MD, Finkelstein EA, Hong Y, Johnston SC, Khera A, Lloyd-Jones DM, Nelson SA, Nichol G, Orenstein D, Wilson PW, Woo YJ (2011) Forecasting the future of cardiovascular disease in the United States: a policy statement from the American Heart Association. Circulation 123(8):933–944

Hershberger RE, Siegfried JD (2011) Update 2011: clinical and genetic issues in familial dilated cardiomyopathy. J Am Coll Cardiol 57(16):1641–1649

Hershberger RE, Morales A, Siegfried JD (2010) Clinical and genetic issues in dilated cardiomyopathy: a review for genetics professionals. Genet Med 12(11):655–667

Hu S, Wilson KD, Ghosh Z, Han L, Wang Y, Lan F, Ransohoff KJ, Burridge P, Wu JC (2013) MicroRNA-302 increases reprogramming efficiency via repression of NR2F2. Stem Cells 31(2):259–268

Jessup M, Abraham WT, Casey DE, Feldman AM, Francis GS, Ganiats TG, Konstam MA, Mancini DM, Rahko PS, Silver MA, Stevenson LW, Yancy CW (2009) 2009 focused update: ACCF/AHA Guidelines for the Diagnosis and Management of Heart Failure in Adults: a report of the American College of Cardiology Foundation/American Heart Association Task Force on Practice Guidelines: developed in collaboration with the International Society for Heart and Lung Transplantation. Circulation 119(14):1977–2016

Jung CB, Moretti A, Schnitzler MM, Iop L, Storch U, Bellin M, Dorn T, Ruppenthal S, Pfeiffer S, Goedel A, Dirschinger RJ, Seyfarth M, Lam JT, Sinnecker D, Gudermann T, Lipp P, Laugwitz KL (2012) Dantrolene rescues arrhythmogenic RYR2 defect in a patient-specific stem cell model of catecholaminergic polymorphic ventricular tachycardia. EMBO Mol Med 4(3):180–191

Kattman SJ, Witty AD, Gagliardi M, Dubois NC, Niapour M, Hotta A, Ellis J, Keller G (2011) Stage-specific optimization of activin/nodal and BMP signaling promotes cardiac differentiation of mouse and human pluripotent stem cell lines. Cell Stem Cell 8(2):228–240

Kushwaha SS, Fallon JT, Fuster V (1997) Restrictive cardiomyopathy. N Engl J Med 336(4): 267–276

Lakdawala NK, Funke BH, Baxter S, Cirino AL, Roberts AE, Judge DP, Johnson N, Mendelsohn NJ, Morel C, Care M, Chung WK, Jones C, Psychogios A, Duffy E, Rehm HL, White E, Seidman JG, Seidman CE, Ho CY (2012) Genetic testing for dilated cardiomyopathy in clinical practice. J Card Fail 18(4):296–303

Lenfant C (2013) Prospects of personalized medicine in cardiovascular diseases. Metabolism 62(Suppl 1):S6–S10

Liang P, Lan F, Lee AS, Gong T, Sanchez-Freire V, Wang Y, Diecke S, Sallam K, Knowles JW, Nguyen PK, Wang PJ, Bers DM, Robbins RC, Wu JC (2013) Drug screening using a library of human induced pluripotent stem cell-derived cardiomyocytes reveals disease specific patterns of cardiotoxicity. Circulation 127(16):1677–1691

Lundy SD, Zhu WZ, Regnier M, Laflamme M (2013) Structural and functional maturation of cardiomyocytes derived from human pluripotent stem cells. Stem Cells Dev 22(14):1991–2002

Maass A, Leinwand LA (2000) Animal models of hypertrophic cardiomyopathy. Curr Opin Cardiol 15(3):189–196

Maron BJ, Maron MS (2013) Hypertrophic cardiomyopathy. Lancet 381(9862):242–255

Maron BJ, Towbin JA, Thiene G, Antzelevitch C, Corrado D, Arnett D, Moss AJ, Seidman CE, Young JB (2006) Contemporary definitions and classification of the cardiomyopathies: an American Heart Association Scientific Statement from the Council on Clinical Cardiology, Heart Failure and Transplantation Committee; Quality of Care and Outcomes Research and Functional Genomics and Translational Biology Interdisciplinary Working Groups; and Council on Epidemiology and Prevention. Circulation 113(14):1807–1816

Maron BJ, Maron MS, Semsarian C (2012) Genetics of hypertrophic cardiomyopathy after 20 years: clinical perspectives. J Am Coll Cardiol 60(8):705–715

Matsa E, Denning C (2012) In vitro uses of human pluripotent stem cell-derived cardiomyocytes. J Cardiovasc Transl Res 5(5):581–592

Matsa E, Rajamohan D, Dick E, Young L, Mellor I, Staniforth A, Denning C (2011) Drug evaluation in cardiomyocytes derived from human induced pluripotent stem cells carrying a long QT syndrome type 2 mutation. Eur Heart J 32(8):952–962

Matsa E, Dixon JE, Medway C, Georgiou O, Patel MJ, Morgan K, Kemp PJ, Staniforth A, Mellor I, Denning C (2014) Allele-specific RNA interference rescues the long-QT syndrome phenotype in human-induced pluripotency stem cell cardiomyocytes. Eur Heart J 35(16): 1078–1087

McNally EM, Golbus JR, Puckelwartz MJ (2013) Genetic mutations and mechanisms in dilated cardiomyopathy. J Clin Invest 123(1):19–26

Morano I (1999) Tuning the human heart molecular motors by myosin light chains. J Mol Med (Berl) 77(7):544–555

Mordwinkin NM, Burridge PW, Wu JC (2013) A review of human pluripotent stem cell-derived cardiomyocytes for high-throughput drug discovery, cardiotoxicity screening, and publication standards. J Cardiovasc Transl Res 6(1):22–30

Moric-Janiszewska E, Markiewicz-Loskot G (2007) Review on the genetics of arrhythmogenic right ventricular dysplasia. Europace 9(5):259–266

Mummery C, Ward-van Oostwaard D, Doevendans P, Spijker R, van den Brink S, Hassink R, van der Heyden M, Opthof T, Pera M, de la Riviere AB, Passier R, Tertoolen L (2003) Differentiation of human embryonic stem cells to cardiomyocytes: role of coculture with visceral endoderm-like cells. Circulation 107(21):2733–2740

Navarrete EG, Liang P, Lan F, Sanchez-Freire V, Simmons C, Gong T, Sharma A, Burridge P, Patlolla B, Lee A, Wu H, Beygui R, Wu S, Robbins RC, Bers D, Wu JC (2013) Screening adverse drug-induced arrhythmia events using human induced pluripotent stem cell-derived cardiomyocytes and low-impedance microelectrode arrays. Circulation 128(11 Suppl 1): S3–S13

Otsuji TG, Minami I, Kurose Y, Yamauchi K, Tada M, Nakatsuji N (2010) Progressive maturation in contracting cardiomyocytes derived from human embryonic stem cells: qualitative effects on electrophysiological responses to drugs. Stem Cell Res 4(3):201–213

Parry HM, Doney AS, Palmer CN, Lang CC (2013) State of play of pharmacogenetics and personalized medicine in heart failure. Cardiovasc Ther 31(6):315–322

Pilichou K, Bezzina CR, Thiene G, Basso C (2011) Arrhythmogenic cardiomyopathy: transgenic animal models provide novel insights into disease pathobiology. Circ Cardiovasc Genet 4(3): 318–326

Rajala K, Pekkanen-Mattila M, Aalto-Setala K (2011) Cardiac differentiation of pluripotent stem cells. Stem Cells Int 2011:383709

Rajamohan D, Matsa E, Kalra S, Crutchley J, Patel A, George V, Denning C (2013) Current status of drug screening and disease modelling in human pluripotent stem cells. Bioessays 35(3): 281–298

Recchia FA, Lionetti V (2007) Animal models of dilated cardiomyopathy for translational research. Vet Res Commun 31(Suppl 1):35–41

Reubinoff BE, Pera MF, Fong CY, Trounson A, Bongso A (2000) Embryonic stem cell lines from human blastocysts: somatic differentiation in vitro. Nat Biotechnol 18(4):399–404

Richardson P, McKenna W, Bristow M, Maisch B, Mautner B, O'Connell J, Olsen E, Thiene G, Goodwin J, Gyarfas I, Martin I, Nordet P (1996) Report of the 1995 World Health Organization/ International Society and Federation of Cardiology Task Force on the Definition and Classification of cardiomyopathies. Circulation 93(5):841–842

Sen-Chowdhry S, Syrris P, McKenna WJ (2010) Genetics of restrictive cardiomyopathy. Heart Fail Clin 6(2):179–186

Soldner F, Jaenisch R (2012) iPSC disease modeling. Science 338(6111):1155–1156

Steeds RP (2013) Multimodality imaging in heart failure patients. Curr Opin Cardiol 28(2): 209–215

Sun N, Yazawa M, Liu J, Han L, Sanchez-Freire V, Abilez OJ, Navarrete EG, Hu S, Wang L, Lee A, Pavlovic A, Lin S, Chen R, Hajjar RJ, Snyder MP, Dolmetsch RE, Butte MJ, Ashley EA, Longaker MT, Robbins RC, Wu JC (2012) Patient-specific induced pluripotent stem cells as a model for familial dilated cardiomyopathy. Sci Transl Med 4(130):130ra47

Takahashi K, Okita K, Nakagawa M, Yamanaka S (2007) Induction of pluripotent stem cells from fibroblast cultures. Nat Protoc 2(12):3081–3089

Van Driest SL, Ommen SR, Tajik AJ, Gersh BJ, Ackerman MJ (2005) Yield of genetic testing in hypertrophic cardiomyopathy. Mayo Clin Proc 80(6):739–744

Wang Y, Zhang WY, Hu S, Lan F, Lee AS, Huber B, Lisowski L, Liang P, Huang M, de Almeida PE, Won JH, Sun N, Robbins RC, Kay MA, Urnov FD, Wu JC (2012) Genome editing of human embryonic stem cells and induced pluripotent stem cells with zinc finger nucleases for cellular imaging. Circ Res 111(12):1494–1503

Xu C, Police S, Hassanipour M, Li Y, Chen Y, Priest C, O'Sullivan C, Laflamme MA, Zhu WZ, Van Biber B, Hegerova L, Yang J, Delavan-Boorsma K, Davies A, Lebkowski J, Gold JD (2011) Efficient generation and cryopreservation of cardiomyocytes derived from human embryonic stem cells. Regen Med 6(1):53–66

Zhang J, Wilson GF, Soerens AG, Koonce CH, Yu J, Palecek SP, Thomson JA, Kamp TJ (2009) Functional cardiomyocytes derived from human induced pluripotent stem cells. Circ Res 104(4):e30–e41

Modeling Arrhythmogenic Right Ventricular Dysplasia/Cardiomyopathy with Patient-Specific iPSCs

3

K. Shah, C.-Y. Wei, C.-S. Kim, J. Wong, J.-Y. Wen, T. Tirasawasdichai, C. Wang, and H.-S. Vincent Chen

Abstract

Cellular reprogramming of somatic cells from cardiac patients to induced pluripotent stem cells (iPSCs) enables in vitro modeling of human genetic disorders for pathogenic investigations and therapeutic screens. However, using iPSC-derived cardiomyocytes (iPSC-CMs) to model an adult-onset heart disease remains challenging due to the uncertainty regarding the ability of relatively immature iPSC-CMs to fully recapitulate adult disease phenotypes. Arrhythmogenic right ventricular dysplasia (ARVD) is an inherited cardiomyopathy characterized by pathological fibrofatty infiltration and cardiomyocyte (CM) loss predominantly in the right ventricle (RV), leading to life-threatening ventricular arrhythmias. Over 50 % of affected individuals have desmosome gene mutations, most commonly in *PKP2* encoding plakophilin-2. The median age at presentation of ARVD is 26–30 years. We used Yamanaka's pluripotent factors to generate iPSC lines from two ARVD patients with *PKP2* mutations. We first developed a method to induce metabolic maturation of iPSC-CMs and showed that induction of adult-like/fatty acid dominant energetics from an embryonic/glycolytic state is essential to model an adult-onset cardiac disease

K. Shah • C.-Y. Wei • J. Wong • J.-Y. Wen • T. Tirasawasdichai • C. Wang
Development, Aging and Regeneration (DARe) Program, Sanford-Burnham-Prebys Medical Discovery Institute, 10901 North Torrey Pines Road, La Jolla, CA 92037, USA

C.-S. Kim
Development, Aging and Regeneration (DARe) Program, Sanford-Burnham-Prebys Medical Discovery Institute, 10901 North Torrey Pines Road, La Jolla, CA 92037, USA

Department of Bioscience and Biotechnology, Sejong University, Gwangjin-Gu, Seoul, Korea

H.-S.V. Chen (✉)
Development, Aging and Regeneration (DARe) Program, Sanford-Burnham-Prebys Medical Discovery Institute, 10901 North Torrey Pines Road, La Jolla, CA 92037, USA

Department of Medicine/Cardiology, University of California-San Diego, La Jolla, CA, USA
e-mail: hsv_chen@sanfordburnham.org

K. Fukuda (ed.), *Human iPS Cells in Disease Modelling*,
DOI 10.1007/978-4-431-55966-5_3

using patient-specific iPSC-CMs. Furthermore, we demonstrate that coactivation of normal peroxisome proliferator-activated receptor-alpha (PPARα) and abnormal PPARγ pathways led to aggressive lipogenesis, elevated apoptosis, and defective intracellular calcium handling in ARVD iPSC-CMs, recapitulating the pathological signatures of ARVD. PPARγ antagonists rescued all ARVD pathological phenotypes and reactive oxygen species (ROS) scavengers curtailed CM apoptosis in our ARVD in vitro model. Thus, using this model, we revealed novel pathogenic insights that metabolic derangement in an adult-like metabolic milieu underlies ARVD pathologies, enabling us to test novel disease-modifying therapeutic strategies.

Keywords
Induced pluripotent stem cell-derived cardiomyocytes • Fatty acid oxidation (FAO) • Peroxisome proliferator-activated receptor (PPAR) • Reactive oxygen species (ROS) • Metabolic maturation

3.1 Introduction

Cardiovascular diseases remain the major cause of death in the world (Go et al. 2014). Recent advances in cellular reprogramming of somatic cells (Takahashi et al. 2007; Yu et al. 2007; Park et al. 2008) from patients with inherited heart diseases into induced pluripotent stem cells (iPSCs) have enabled the generation of cardiomyocytes (CMs) for myocardial repair (Okano et al. 2013; Shiba et al. 2012; Murry et al. 2006; Chen et al. 2009) and in vitro modeling of human-inherited cardiac diseases (Carvajal-Vergara et al. 2010; Moretti et al. 2010, 2013; Itzhaki et al. 2011; Yazawa et al. 2011; Kim et al. 2013; Shinnawi and Gepstein 2014; Mercola et al. 2013). However, CMs derived from human embryonic stem cells or iPSCs (hESC-CMs or iPSC-CMs) displayed heterogeneous and immature phenotypes (Kim et al. 2010; Veerman et al. 2015; Yang et al. 2014). In fact, most published iPSC-based cardiac disease models used immature patient-specific iPSC-CMs. Rapid-onset pathologies and exaggerated arrhythmias with these iPSC-based models frequently occurred spontaneously within 30 days in culture, which deviated greatly from the clinical course of these cardiac diseases (Veerman et al. 2015; Yang et al. 2014; Knollmann 2013; Santostefano et al. 2015). Thus, there is a tremendous need to develop methods that induce maturation of primitive iPSC-CMs so that better and more clinically relevant cardiac disease models can be established for pathogenic and therapeutic investigations.

3.2 Pathologies and Pathogenic Mechanisms of Arrhythmogenic Right Ventricular Dysplasia/ Cardiomyopathy (ARVD/C)

ARVD/C (termed ARVD here for simplicity) is an inherited cardiomyopathy with most identified mutations in genes that encode cardiac desmosomes, which include plakoglobin (*JUP*), plakophilin-2 (*PKP2*), desmoplakin (*DSP*), desmoglein-2 (*DSG2*), and desmocollin-2 (*DSC2*) (Marcus et al. 1982; Calkins 2015; den Haan et al. 2009; Awad et al. 2008; Basso et al. 2012). Pathological hallmarks of ARVD are progressive fibrofatty replacement of CMs with increased CM apoptosis primarily in the RV, leading to lethal arrhythmias. Clinical criteria to diagnose ARVD has been well established (Marcus et al. 2010). Pathogenic mechanisms of ARVD are difficult to study because obtaining cardiac samples from early stages of human ARVD hearts is rarely possible due to ARVD being commonly diagnosed at advance diseased stages or postmortem. Additionally, primary cardiac tissues are difficult to biopsy safely from symptomatic ARVD patients due to the risk of cardiac perforation. These limiting factors impose significant constraints in developing therapies for human ARVD. Currently, no therapy is available to stop the progression of ARVD pathologies except inserting implantable cardioverter defibrillators (ICDs) to prevent sudden cardiac death (Calkins 2015).

The pathogenic mechanisms by which desmosomal mutations cause CM loss, fibrofatty infiltration, and lethal arrhythmia remain poorly understood. Experimental data from animal and cultured cell line models have led to confusing and conflicting results (Garcia-Gras et al. 2006; Asimaki et al. 2009; Basso et al. 2013; Li et al. 2011; Swope et al. 2012; Zhang et al. 2015). Early experiments with *DSP* knockdown in murine HL-1 atrial tumor CMs and cardiac-specific *DSP* knockout (KO) mice suggested that *DSP* deficits resulted in aggressive lipogenesis in CMs and nuclear translocation of Plakoglobin proteins (Pkg). This abnormal Pkg nuclear localization was suggested to compete with and decrease the binding of β-catenin to the TCF/LEF transcription factor complexes (gain of function hypothesis), leading to low β-catenin activities and adipogenic transdifferentiation of CMs (Garcia-Gras et al. 2006). However, the majority of human ARVD heart tissues demonstrated significant downregulation of Pkg without abnormal nuclear translocation (Asimaki et al. 2009; Basso et al. 2013), and cardiac-specific KO of Pkg can reproduce ARVD pathologies in mouse hearts (Li et al. 2011; Swope et al. 2012), indicating that loss of function in Pkg, rather than Pkg competition with β-catenin, is responsible for eliciting ARVD pathologies (Zhang et al. 2015).

Moreover, using genetic fate mapping techniques in mouse ARVD models, Islet-1-positive (Isl1+) cardiac progenitor cells of the second heart field have been implicated as the source of adipocytes in ARVD hearts (Lombardi et al. 2009). However, the incorrect use of activation of PPARγ and its target genes as the markers for "adipogenesis (adipocyte formation)" likely accounted for this misleading interpretation (Garcia-Gras et al. 2006; Lombardi et al. 2009). In fact, PPARγ and its target genes are normally activated in several non-adipocyte tissues (e.g. endothelial cells and macrophages) (Willson et al. 2001) and could be abnormally activated by pathological conditions in hepatocytes (obesity) (Pettinelli and Videla 2011) and CMs (diabetic cardiomyopathy, Marfella et al. 2009). Therefore, activation of PPARγ and its target genes indicates the active "de novo lipogenesis (DNL)" rather than cell transdifferentiation toward adipocytes. More importantly, the proper lineage marker for identifying adipocytes is perilipin-1 (*PLIN1*) that is only expressed in adipocytes (Brasaemle 2007).

3.3 Establishing iPSCs from Patients with Clinical ARVD and *PKP2* Mutations

To further explore the pathogenic mechanisms of human ARVD hearts, we generated and fully characterized two sets of ARVD *PKP2* mutant iPSC lines from two unrelated ARVD patients (Kim et al. 2013). The first ARVD patient has homozygous c.2484C>T mutations in *PKP2* proteins (Pkp2) with frame-shifted C-terminals failing to anchor Pkg to the sarcolemmal membrane (JK lines) (Awad et al. 2006). The second patient has a heterozygous c.2013delC mutation in *PKP2* (termed delC *PKP2* mutation) (Dalal et al. 2006) that causes frameshift and premature termination in exon 10, leading to the degradation of mutant *PKP2* mRNAs and a *PKP2* haploinsufficiency phenotype. Importantly, we observed abnormal nuclear translocation of Pkg (only in a monolayer format, Fig. 3.1a) and very low β-catenin activity in both *PKP2* mutant ARVD iPSC-CMs at baseline, but no exaggerated lipogenesis or apoptosis in ARVD iPSC-CMs was found after culturing for 2–3 months (Kim et al. 2013), indicating that Pkg nuclear translocation and low β-catenin activities are insufficient for generating ARVD pathologies. Also, complete Pkg nuclear translocation occurred only in the ARVD iPSC-CMs contacting the rigid plastic culture surface (>6 gigapascal) but not in the center of beating clusters with physiological stiffness (Fig. 3.1), indicating that dominant Pkg nuclear localization is a culture-induced phenomenon and not a clinically relevant pathological finding (Asimaki et al. 2009; Basso et al. 2013). Moreover, we did not observe any CM transdifferentiation to adipocytes in our Isl1+ cardiac progenitor cell populations derived from human ARVD iPSCs in any culture condition. These results are consistent with the loss of function role of Pkg in the pathogenesis of ARVD hearts (Zhang et al. 2015).

3.4 Designing the Method of Metabolic Maturation Induction

The major metabolic differences between embryonic and adult CMs are (1) embryonic CMs use mainly glycolysis and lactate for energy production and (2) adult CMs produce most energy via fatty acid oxidation (FAO) but retain the ability to readily switch to glucose or other substrate utilization when fatty acid is not available or FAO is compromised (Onay-Besikci 2006; Lopaschuk et al. 2010). At baseline, the potential energy substrates in our culture media (DMEM) are 25 mM D-glucose, 1 mM Na^+ pyruvate, 2 mM glutamax, essential amino acids,

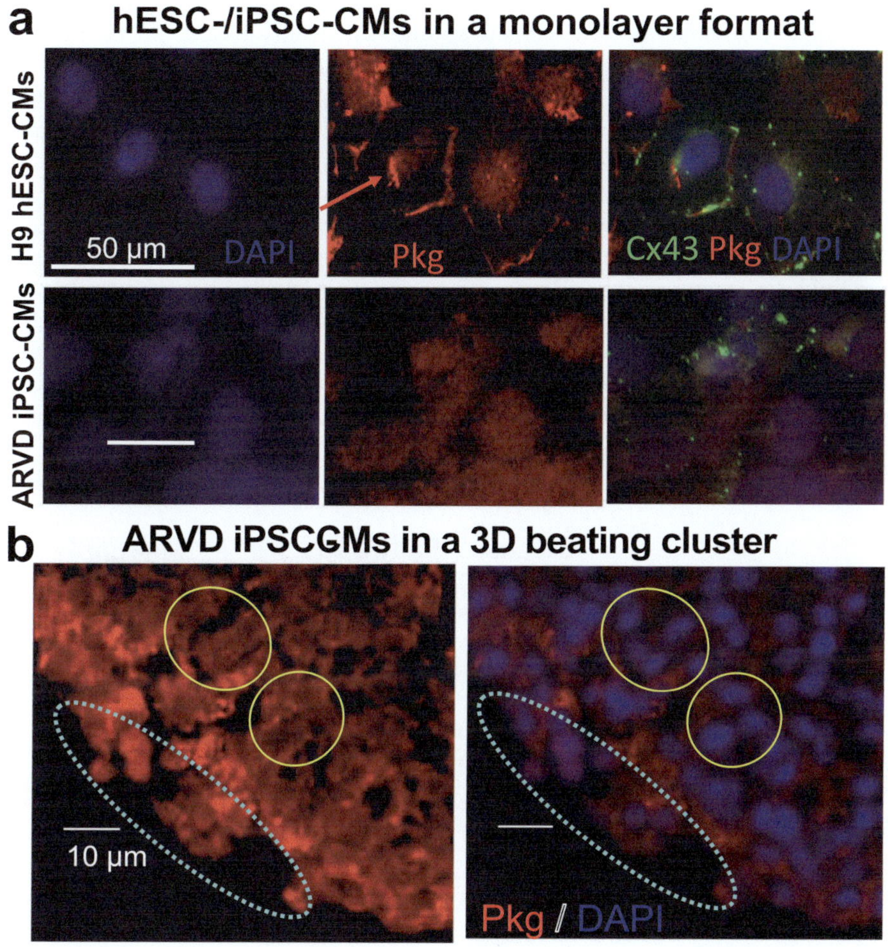

Fig. 3.1 Stiffness of culture substrates affects the Pkg distribution. (**a**) In a monolayer format on stiff plastic surfaces, both normal H9 hESC-CMs and ARDV iPSC-CMs displayed significant nuclear Pkg staining. (**b**) In a 3D beating cluster, strong Pkg nuclear staining in ARVD iPSC-CMs only occurred at the cluster edges with contact to the plastic surface (*dotted yellow circle*) but not at the center of cluster with cardiac stiffness (*solid yellow circles*). *Red arrow*: membrane Pkg

1 mM NEAA, lipids from de novo lipogenesis, and minimal energy substrates contained in the 2 % fetal bovine serum (FBS). This kind of culture media with high glucose content is very standard for growing human ESC-/iPSC-CMs (Carvajal-Vergara et al. 2010; Moretti et al. 2010; Itzhaki et al. 2011; Yazawa et al. 2011; Kim et al. 2010, 2013). Also, based on our metabolic assays (Kim et al. 2013), both normal and ARVD iPSC-CMs at baseline showed low oxygen consumption rates (OCR) from FAO but high levels of glycolysis by the extracellular acidification rate

(ECAR) measurement, indicating that glycolysis is the main mechanism of energy production for immature iPSC-CMs in culture (see details in Sect. 3.7 below).

We then used insulin, a glucocorticoid [dexamethasone (Dexa)], and 3-isobutyl-1-methylxanthine (IBMX, a phosphodiesterase inhibitor) to develop the three-factor (3F) protocol to drive metabolic maturation (Fig. 3.2). Among these three factors, glucocorticoid is catabolic and can increase mitochondria oxidation of glucose and fatty acids in muscles but promote both lipogenesis and lipolysis (triacylglycerol turnover) in adipocytes (Macfarlane et al. 2008). Insulin is known to inhibit FAO but promote glucose utilization, endogenous de novo lipogenesis, and lipid storage in various cell types (Gregoire et al. 1998). Also, increased cyclic AMP by IBMX promotes both lipogenesis and lipolysis and stimulates FAO in cardiac muscles (Lopaschuk et al. 2010). Increased de novo lipogenesis would produce distinct phosphatidylcholine species that can act as the endogenous ligand to specifically activate PPARα in liver (Chakravarthy et al. 2009) or in cardiomyocytes (Razani et al. 2011). For these aforementioned reasons, we

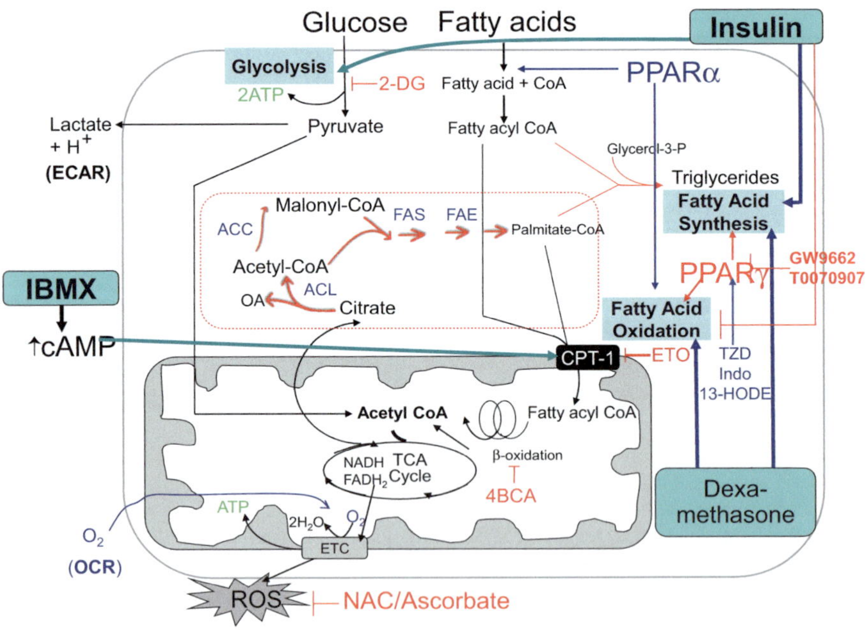

Fig. 3.2 Rationales for designing metabolic maturation induction of primitive iPSC-CMs. Insulin increases glucose utilization and de novo lipogenesis but inhibits FAO. IBMX increases cAMP levels that in turn improve FAO. Dexamethasone increases both FAO and lipogenesis. *Blue arrows* denote normal increase of function; *red arrows*, abnormal increase of function; and *blocked bars*, inhibition of function. *2-DG* denotes 2-deoxyglucose, *CPT-1* carnitine palmitoyl-transferase-1, *TCA* the tricarboxylic acid or Krebs cycle, *ETC* electron transport chain, *TZD* thiazolidinediones, *Indo* indomethacin, *ACL* ATP citrate lyase, *ACC* acetyl-CoA carboxylase, *FAS* fatty acid synthase, *FAE* fatty acid elongase, and *OA* oxaloacetate

designed the 3F protocol to activate PPARα and increase lipogenesis and FAO in iPSC-CMs (Kim et al. 2013). Our functional assay by the Seahorse metabolic flux analyzer (without added external fats) demonstrated that both ARVD and normal iPSC-CMs after 3F treatment had higher absolute OCRs from both fatty acid and glucose oxidation than un-stimulated iPSC-CMs at 60 days after differentiation (60D), supporting that our 3F protocol indeed induced mitochondrial maturation with active oxidative phosphorylation (OXPHOS) (Kim et al. 2013). No significant apoptosis was found in normal and ARVD iPSC-CMs after 3F induction. Thus, activation of PPARα alone by 3F in iPSC-CMs induces an adult-like metabolism without provoking any ARVD pathology (Fig. 3.3).

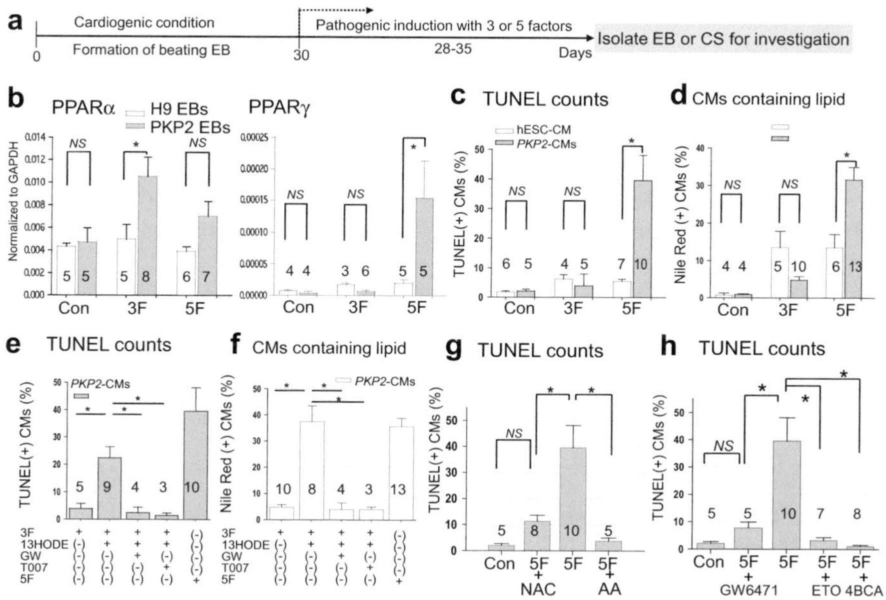

Fig. 3.3 Metabolic maturation and pathogenic induction of ARVD iPSC-CMs and potential therapeutic strategies. (**a**) Protocols for metabolic maturation (3F) and pathogenic (5F) induction. (**b**) 3F-induced PPARα expression and 5F-induced abnormal PPARγ activation only in ARVD iPSC-CMs (relative to GAPDH). (**c**) Degrees of CM apoptosis (TUNEL-positive) and (**d**) percentage of lipid-laden (Nile-red positive) CMs after 0F (Con), 3F, or 5F treatment for 4 weeks. Summary of (**e**) degrees of apoptosis and (**f**) lipid-laden CMs in beating EBs is shown here. 13-HODE can replace the indomethacin and rosiglitazone in the 5F protocol to induce ARVD pathologies, which could be prevented by PPARγ antagonists, GW9662 (GW) or T0070907 (T007); ROS scavengers, NAC or AA (**g**); or scaling down FAO by etomoxir (ETO) or 4-BCA (**h**). The number in each column represents the number of biological replicates tested. *Asterisks* indicate $p < 0.05$ and *NS* no significant difference by ANOVA. All data are shown as mean ± s.e.m (This figure is adapted from Kim et al. 2013)

3.5 Metabolic Maturation-Based Pathogenic Induction of ARVD Pathologies

PPARα is the major transcriptional regulator of fatty acid metabolism in adult cardiomyocytes (Lopaschuk et al. 2010). In contrast, PPARγ should be minimally activated in normal cardiomyocytes. However, PPARγ pathway has been reported to be abnormally hyperactivated in RV tissue samples of ARVD hearts (Djouadi et al. 2009), but transgene-induced overexpression of PPARγ in mouse cardiomyocytes could lead to dilated cardiomyopathy (Son et al. 2007). Based on these two lines of evidence, we added two PPARγ activators, 5 μM rosiglitazone (Willson et al. 2001) and 200 μM indomethacin (Waku et al. 2010), to the 3F protocol to further induce abnormal PPARγ activation in beating embryoid bodies (EBs) [termed the 5-factor (5F) protocol, Fig. 3.2]. We found that mutant *PKP2* iPSC-CMs manifested ARVD pathologies only after induction of adult-like metabolism and abnormal activation of the PPARγ pathway. Coactivation of normal PPARα and abnormal PPARγ pathways resulted in exaggerated lipogenesis, apoptosis, further Na^+ channel downregulation (Kim et al. 2013; Cerrone et al. 2014), and defective intracellular calcium (Ca^{2+}) handling capability in ARVD iPSC-CMs, recapitulating the pathological signatures of ARVD (Fig. 3.3b–d, and see Kim et al. 2013). We further showed that the endogenous PPARγ activator 13-hydroxyocta-decadienoic acid [13-HODE, a major component of the oxidized low-density lipoprotein (oxLDL)] (Willson et al. 2001), could replace rosiglitazone and indomethacin in the 5F protocol for abnormal PPARγ activation, leading to extensive lipogenesis and apoptosis in mutant iPSC-CMs (Fig. 3.3e–f). Importantly, overexpression of wild-type (WT) *PKP2* in ARVD iPSC-CMs reversed 5F-induced ARVD pathologies (Kim et al. 2013), supporting that observed pathological phenotypes are the result of mutated *PKP2* genes in iPSCs.

3.6 Potential Therapeutic Strategies for Treating ARVD Pathologies

We first showed that PPARγ antagonists, e.g., GW9662 (GW, Willson et al. 2001) or T0070907 (T007, Waku et al. 2010), rescued all ARVD pathologies (Fig. 3.3e–f) and ROS scavengers [N-acetyl-cysteine (NAC) or ascorbic acid (AA)] curtailed CM apoptosis in our ARVD in vitro model (Fig. 3.3g). Of note, activation of PPARγ alone by rosiglitazone and indomethacin or blockade of PPARα-dependent metabolism alone with an antagonist (GW6471) during 5F treatment did not induce significant ARVD pathologies, supporting the crucial role of PPARα/PPARγ coactivation in mediating ARVD pathogenesis.

Moreover, we observed that the absolute levels of FAO in ARVD iPSC-CMs are always 1.5 to 2-fold higher than those of normal iPSC-CMs after 3F metabolic maturation induction (Fig. 3.4). This higher FAO flux in ARVD iPSC-CMs might explain why many patients with desmosomal mutations participated and excelled in endurance sports at younger ages prior to the pathological processes kicked in at adulthood. However, endurance exercises are known to accelerate the pathological phenotypes of ARVD (Sawant and Calkins 2015). The elevated cAMP levels from the continuous presence of IBMX in our 3F or 5F protocols likely simulate endurance exercise in the real life. We therefore hypothesized that this high FAO flux might also account for the high ROS production in ARVD iPSC-CMs and is required to drive them toward the apoptotic state after PPARα/PPARγ coactivation. As such, we toned down FAO hyper-flux by adding an EC50 dose of either etomoxir (ETO, to block CPT-1) or 4-bromocrotonic acid (4BCA, to block β-oxidation), which reduced CM apoptosis (Fig. 3.3h). This result strongly supports the use of beta-blockers and the recommendation of avoidance of endurance/competitive sports (Marcus et al. 2010) so as to reduce the metabolic flux and the progression of ARVD phenotypes.

3.7 Distinct Metabolic Phenotypes of Immature, Mature, and Pathological ARVD iPSC-CMs

Using the Seahorse XF96 Extracellular Flux Analyzer (Ferrick et al. 2008), functional assays of FAO and glycolysis in live cells revealed that normal and ARVD iPSC-CMs had dominant glycolytic energetics (an embryonic pattern) at the baseline (Fig. 3.4a–h). After activation of PPARα by 3F, mutant and normal iPSC-CMs displayed slightly higher levels of glycolysis but significant activation of FAO (an adult-like pattern) when compared to the un-induced, baseline conditions [zero factor (0F)]. Compared to 3F induction, mutant *PKP2* iPSC-CMs treated with 5F for 4 weeks demonstrated overall depressed energy metabolism with more FAO reduction than the reduction in glycolysis, resulting in a fuel shift from using both fatty acids and glucose to primarily glucose utilization (including glycolysis and pyruvate oxidation), much like the so-called metabolic burnt-out state observed in failing hearts (Neubauer 2007). Pictures of 0F- and 5F-treated ARVD iPSC-CMs by electron microscopy (EM) further demonstrated mature mitochondria with long and dense cristae (Fig. 3.5) when compared to immature mitochondria with short and sparse cristae in the immature iPSC-CMs. Thus, results from immunocytochemical, genetic, EM, and metabolic assays strongly support the notion that coactivation of PPARα and PPARγ by hormones and small molecules accelerate the pathogenesis and establish an efficient in vitro model of ARVD, recapitulating the metabolic and pathological signatures of failing ARVD hearts within 2 months.

3.8 Calcium Handing Deficits and Cardiac Sodium (Na^+) Channel Downregulation in ARVD iPSC-CMs

The clinical lethal arrhythmias in ARVD patients are sustained, and reentrant ventricular tachycardia (VT) that could not be modeled without a 3D multilayer cardiac system (Calkins 2015). However, the burden of premature ventricular contractions appears to correlate with the frequency of clinical VT (Calkins 2015). In order to

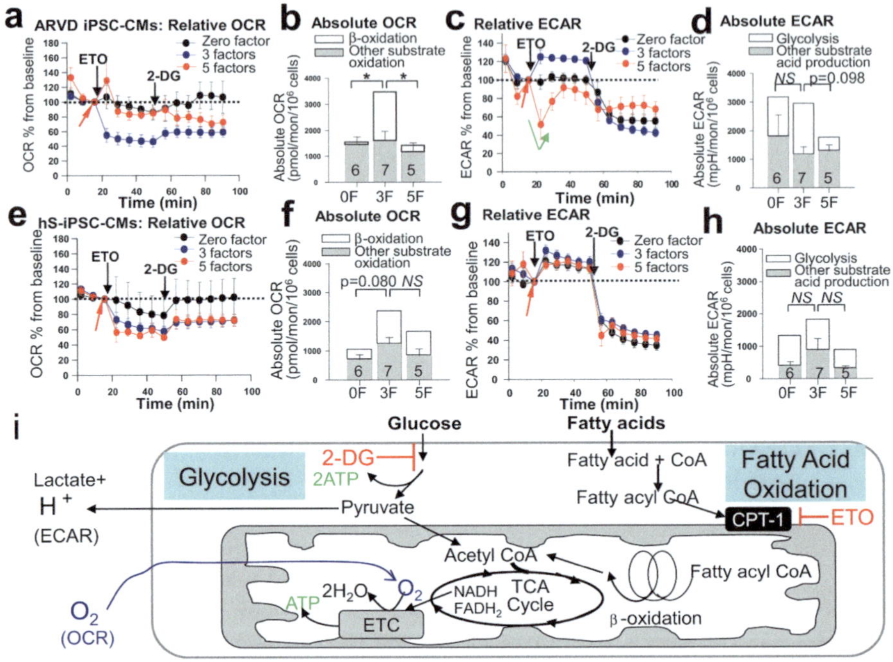

Fig. 3.4 Functional assays of glucose utilization versus FAO in *PKP2* mutant iPSC-CMs after various treatments. Using a Seahorse XF96 Extracellular Flux Analyzer, we determined the degree of FAO by measuring the etomoxir (100 μM ETO, a specific CPT-1 inhibitor) blocked components of OCR, and glycolysis by measuring 2-deoxyglucose (50 mM 2-DG) blocked components of ECAR (see Ref Chen et al. (2009) for detailed methods). Absolute values of OCR and ECAR are expressed as pmol/min/10^6 cells and mpH/min/10^6 cells, respectively. (**a**) Real-time measurement of OCR showed that ETO-blocked 14.4 ± 10.2 (0F), 53.9 ± 8.2 (3F), and 15.3 ± 8.5 % (5F) of baseline OCR (*red arrow*) for ARVD iPSC-CMs. (**b**) ETO-blocked absolute OCR (*white boxes*) for ARVD iPSC-CMs are 85.2 ± 196.6 (0F), 1865.3 ± 383.5 (3F), and 254.5 ± 115.4 (5F). (**c**) ECAR measurement after ETO inhibition of β-oxidation showed a rapid ~21 % compensatory increase of glycolysis only in the 3F condition (a switch in energy substrates); yet, after 5F, ETO transiently decreased ECAR by ~45 % (*green arrow*) followed by ~28 % compensatory increase in ECAR (glycolysis). (**d**) 2DG-blocked absolute ECAR (glycolysis) for ARVD iPSC-CMs after 0F, 3F, or 5F are 1353.0 ± 313.6, 1766.0 ± 579.7, and 457.2 ± 211.0, respectively. Comparable patterns in absolute and relative OCR or ECAR in normal hS-iPSC-CMs (WS#4) are shown in (**e–h**). For normal iPSC-CMs, ETO-blocked FAO after 0F, 3F, or 5F are (**e**) (relative) 21.8 ± 24.8 %, 42.6 ± 6.6 %, and 50.5 ± 4.06 % or (**f**) (absolute) 329.4 ± 235.8, 1114.5 ± 316.6, and 807.8 ± 196.4, respectively. 2DG-blocked glycolysis after 0F, 3F, or 5F are (**g**) (relative) 65.5 ± 5.3 %, 54.9 ± 3.8 %, and 58.8 ± 4.8 % or (**h**) (absolute) 915.3 ± 270.6, 929.4 ± 314.3, and 556.8 ± 217.9, respectively. (**i**) A simple diagram to illustrate substrate utilization pathways in CMs. These results (**a–h**) support that both normal and ARVD iPSC-CMs (1) display embryonic metabolism at the baseline and (2) show significantly increased FAO after 3F with ability to switch between FAO and glucose utilization (an adult-like metabolic pattern). ARVD iPSC-CMs after 5F behave like failing CMs with pathological glucose-dominant metabolism. *Single asterisk* indicates $p < 0.05$ and *NS* no significant difference by ANOVA. *P*-values are shown when unpaired *t*-test was performed (This figure is adapted from Kim et al. 2013)

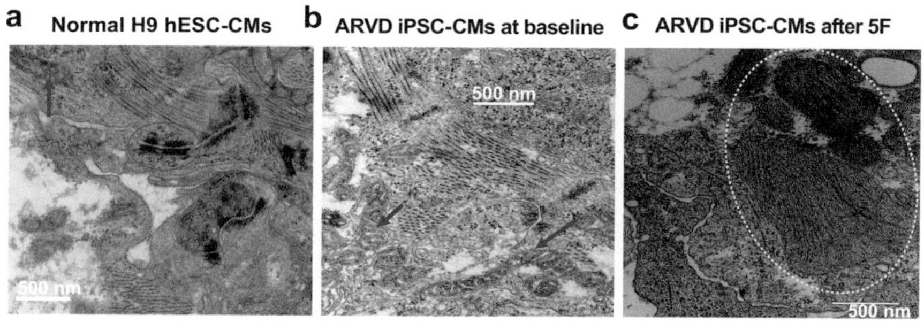

Fig. 3.5 Increased mitochondrial size and content after metabolic induction. EM pictures of (**a**) a 60D H9 hESC-CM at baseline, (**b**) a 60D ARVD iPSC-CM at baseline, and (**c**) an ARVD CM after 5F induction are shown. Normal hESC-CM displays intact desmosomal junctions (*blue circle*) but small, immature mitochondria with short and sparse cristae (*blue arrows*). The ARVD CM after 5F induction shows no visible desmosomal junction, large and mature mitochondria-containing dense and long cristae (*dotted yellow circle*)

further explore the abnormal excitability of ARVD iPSC-CMs, we used Ca^{2+} imaging and intracellular/patch-clamp recordings to evaluate the arrhythmogenic potentials of iPSC-CMs. Compared to normal H9 hESC-CMs at baseline, ARVD iPSC-CMs demonstrated slower intracellular calcium ($[Ca^{2+}]_i$) relaxation after pacing only in high extracellular Ca^{2+} media (5.8 mM) but not in normal Ca^{2+} media (1.8 mM). However, after 5F pathogenic induction, ARVD iPSC-CMs demonstrated a prolonged $[Ca^{2+}]_i$ relaxation phase even in normal Ca^{2+} media after pacing (Fig. 3.6a–b), indicating abnormal $[Ca^{2+}]_i$ handling capability. Furthermore, quantitative-PCR (qPCR) analysis of ARVD iPSC-CMs at baseline showed mildly decreased expression levels of sarcoplasmic reticulum Ca^{2+}-ATPase (SERCA, for Ca^{2+} reuptake) with preserved Na^+/Ca^{2+} exchanger 1 (NCX1, for $[Ca^{2+}]_i$ extrusion) expression, which may explain the impaired $[Ca^{2+}]_i$ relaxation of ARVD iPSC-CMs only in high extracellular Ca^{2+} media at baseline. After 5F treatment, both NCX1 and SERCA expression levels were significantly downregulated relative to hESC-CMs, which would account for the further impaired $[Ca^{2+}]_i$ relaxation phase of ARVD iPSC-CMs even in normal Ca^{2+} media (Kim et al. 2013). Also, cardiac Na^+ channel subunit ($Na_V1.5$) expression and Na^+ currents (INa^+) of ARVD iPSC-CMs are 80 % lower at baseline and show further downregulation after pathogenic induction than the normal hESC-CMs. Future mechanistic studies are needed to determine the roles of impaired $[Ca^{2+}]_i$ handling and altered electrical properties in mediating pathologies or arrhythmia of ARVD iPSC-CMs (Awad et al. 2008).

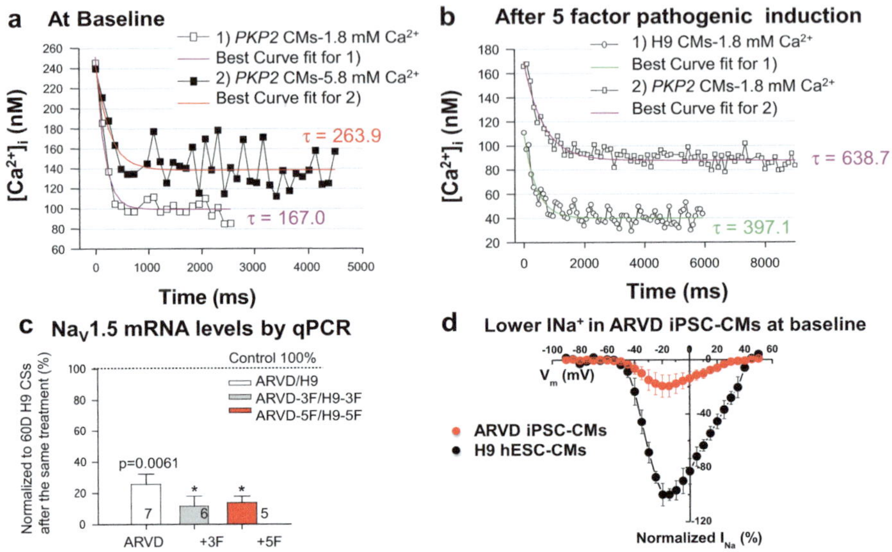

Fig. 3.6 Impaired $[Ca^{2+}]_i$ handling capability and decreased Na^+ currents (INa) in ARVD iPSC-CMs. (**a**) Ca^{2+} imaging with Fura-2AM and single exponential curve fitting of the relaxation phase showed ARVD iPSC-CMs demonstrated prolonged relaxation time constant (τ) only in high Ca^{2+} media (5.8 mM). (**b**) After pathogenic induction with 5F, ARVD iPSC-CMs showed prolonged τ even in the normal Ca^{2+} media (1.8 mM). (**c**) Using qPCR to compare the $Na_V1.5$ mRNA expression levels of ARVD JK#2 and #11 iPSC-CMs to H9 hESC-CMs after comparable treatment, we find that $Na_V1.5$ mRNA transcript is downregulated in ARVD iPSC-CMs and decreases further with 5F pathogenic induction. (**d**) Patch-clamp recordings also show decreased INa in ARVD iPSC-CMs at the baseline when compared to normal H9 hESC-CMs (This figure is adapted from Kim et al. 2013)

3.9 iPSC-CMs Derived from Islet-1+-Rich Second Heart Field Cardiac Progenitor Cells Displayed More Severe ARVD Disease Phenotypes

In order to explain the predominant pathology in RV of ARVD hearts, we increased the number of Islet-l-positive (Isl1+) cardiac progenitor cells by ~fourfold in JK#2 & #11 ARVD iPSCs with 1 μM 6-bromoindirubin-3′-oxime (BIO) (Qyang et al. 2007) to simulate natural RV formation from Isl1+ progenitor cells of the secondary heart field (Cai et al. 2003) (Fig. 3.7). We also used 100 ng/ml Dickkopf-1 (Dkk1) (Qyang et al. 2007) to decrease the number of Isl1+ progenitor cells and, subsequently, RV-like cardiomyocytes. After 5F pathogenic induction, EBs with enriched Isl1+ cells had significantly more lipogenesis and apoptosis in cardiomyocytes than EBs with minimal numbers of Isl1+ progenitor cells (Fig. 3.7). This result supports the notion that Isl1+ cells confer the dominant pathologies in the RV.

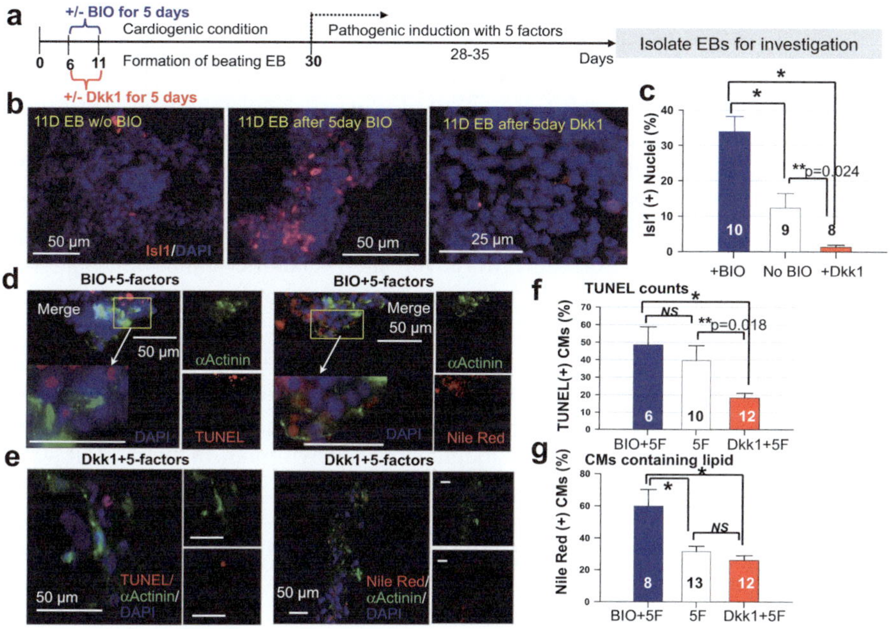

Fig. 3.7 Enrichment of Isl1+ cardiac progenitor cells (CPCs) confers more pathologies in ARVD iPSC-CMs after pathogenic induction. (**a**) Protocols of using BIO to enrich or Dickkopf-1 (Dkk1) to decrease the number of Isl1+ CPCs in ARVD iPSC-EBs (JK#2 and #11 lines). (**b**) Immunolabeling with anti-Isl1 antibodies showed that 5-day treatment of ARVD iPSC-EBs with 1 μM BIO increased the number of Isl1+ CPCs while treatment with 100 ng/ml Dkk1 decreased it. (**c**) BIO or Dkk1 treatment changed the percentage of Isl1+ CPCs from 12.4 ± 4.1 % (baseline) to 34.0 ± 4.2 % or 1.4 ± 0.6 % in 11D EBs, respectively. After (**d**) BIO or (**e**) Dkk1 treatment, 4–5 weeks of 5F pathogenic induction showed much higher CM apoptosis and lipid-laden CMs in BIO-treated than Dkk1-treated ARVD iPSC-CMs in beating EBs. Summary of (**f**) CM apoptosis or (**g**) lipid-laden CMs in Isl1(+)-enriched versus Isl1(+)-poor beating ARVD iPSC-EBs. *Single asterisk* indicates $p < 0.05$ with ANOVA; *double asterisks*, *p*-values significant by unpaired *t*-test only; and *NS* no statistically significant difference

3.10 Conclusions and Future Perspective

Using patient-specific mutant *PKP2* iPSC-CMs grown in a 3D beating EB format and induced by various maturation/pathogenic conditions, we accelerate the pathogenesis of an adult-onset cardiac disease. We demonstrate the importance of PPARα-dependent/adult-like metabolism, PPARγ coactivation, ROS production, and high flux of fatty acid oxidation in the pathogenesis of ARVD. This efficient in vitro iPSC-CM-based model recapitulates the pathognomonic features of ARVD hearts and enables pathogenic investigation and therapeutic screens. Future research to elucidate how mutant Pkp2 leads to abnormal PPARγ activation in ARVD CMs would further deepen our understanding of this unfortunate disease and pave the way for developing novel therapeutics.

Furthermore, we observed only abnormal "lipogenesis" in CMs but not exaggerated adipogenesis in our cardiomyocyte-centric model, suggesting that non-CMs must be involved in the abnormal adipogenesis in ARVD hearts. Also, we did not observe any transdifferentiation of iPSC-CMs to adipocytes (perilipin-1 positive cells). Of note, cardiac fibrosis is also a major component of ARVD hearts. Since human hearts contain a large number of various non-CMs (Chen et al. 2009; Souders et al. 2009), elucidating the subtype(s) of non-CMs that account for the abnormal adipogenesis and/or fibrosis will further improve our in vitro model and render it more clinically relevant. Moreover, the metabolic maturation induction method can be improved further with the addition of several postnatal hormones to achieve true adult levels of metabolic maturation.

Finally, many so-called cardiac disease models have been established by using immature and isolated patient-specific iPSC-CMs (review in Mercola et al. 2013; Moretti et al. 2013; Shinnawi and Gepstein 2014). As a result, rapid-onset pathologies (within 30 days) and spontaneous arrhythmias in these iPSC-CM-based models are frequently observed in these nonphysiological culturing milieus and without any maturation or pathogenic induction. These spontaneous and exaggerated pathologies deviated greatly from the clinical course of these postnatal cardiac disorders, which usually take years to manifest their clinical phenotypes. The clinical relevance of these iPSC-CM-based cardiac disease models is therefore questionable. Thus, we must devote effort and time into studying the mechanisms of cardiomyocyte maturation so as to build better and more clinically relevant in vitro models before an applicable therapeutic screen can be pursued.

Acknowledgments We thank the patients for their participation, microarray core facilities at SBMRI for their support, the Johns-Hopkins ARVD registry for their valuable support, and George W. Rogers from Seahorse Bioscience for assistance in metabolic assays. C-Y. W. is supported by a CIRM training grant (TG2-01162). H-S. V. C. is supported by grants from NIH (RO1 HL105194) and California Institute of Regenerative Medicine (CIRM RB2-01512 & RB4-06276).

Conflicts of Interest The authors report no conflicts of interest in relation to this article.

References

Asimaki A, Tandri H, Huang H, Halushka MK, Gautam S, Basso C, Thiene G, Tsatsopoulou A, Protonotarios N, McKenna WJ, Calkins H, Saffitz JE (2009) A new diagnostic test for arrhythmogenic right ventricular cardiomyopathy. N Engl J Med 360(11):1075–1084

Awad MM, Dalal D, Tichnell C, James C, Tucker A, Abraham T, Spevak PJ, Calkins H, Judge DP (2006) Recessive arrhythmogenic right ventricular dysplasia due to novel cryptic splice mutation in PKP2. Hum Mutat 27(11):1157

Awad MM, Calkins H, Judge DP (2008) Mechanisms of disease: molecular genetics of arrhythmogenic right ventricular dysplasia/cardiomyopathy. Nat Clin Pract Cardiovasc Med 5(5):258–267

Basso C, Bauce B, Corrado D, Thiene G (2012) Pathophysiology of arrhythmogenic cardiomyopathy. Nat Rev Cardiol 9(4):223–233

Basso C, Pilichou K, Thiene G (2013) Is it time for plakoglobin immune-histochemical diagnostic test for arrhythmogenic cardiomyopathy in the routine pathology practice? Cardiovasc Pathol 22(5):312–313

Brasaemle DL (2007) Thematic review series: adipocyte biology. The perilipin family of structural lipid droplet proteins: stabilization of lipid droplets and control of lipolysis. J Lipid Res 48(12):2547–2559

Cai CL, Liang X, Shi Y, Chu PH, Pfaff SL, Chen J, Evans S (2003) Isl1 identifies a cardiac progenitor population that proliferates prior to differentiation and contributes a majority of cells to the heart. Dev Cell 5(6):877–889

Calkins H (2015) Arrhythmogenic right ventricular dysplasia/cardiomyopathy- three decades of progress. Circ J 79(5):901–913

Carvajal-Vergara X, Sevilla A, D'Souza SL, Ang YS, Schaniel C, Lee DF, Yang L, Kaplan AD, Adler ED, Rozov R, Ge Y, Cohen N, Edelmann LJ, Chang B, Waghray A, Su J, Pardo S, Lichtenbelt KD, Tartaglia M, Gelb BD, Lemischka IR (2010) Patient-specific induced pluripotent stem-cell-derived models of LEOPARD syndrome. Nature 465(7299):808–812

Cerrone M, Lin X, Zhang M, Agullo-Pascual E, Pfenniger A, Chkourko Gusky H, Novelli V, Kim C, Tirasawadichai T, Judge DP, Rothenberg E, Chen HS, Napolitano C, Priori SG, Delmar M (2014) Missense mutations in plakophilin-2 cause sodium current deficit and associate with a Brugada syndrome phenotype. Circulation 129(10):1092–1103

Chakravarthy MV, Lodhi IJ, Yin L, Malapaka RR, Xu HE, Turk J, Semenkovich CF (2009) Identification of a physiologically relevant endogenous ligand for PPARalpha in liver. Cell 138(3):476–488

Chen HS, Kim C, Mercola M (2009) Electrophysiological challenges of cell-based myocardial repair. Circulation 120(24):2496–2508

Dalal D, Molin LH, Piccini J, Tichnell C, James C, Bomma C, Prakasa K, Towbin JA, Marcus FI, Spevak PJ, Bluemke DA, Abraham T, Russell SD, Calkins H, Judge DP (2006) Clinical features of arrhythmogenic right ventricular dysplasia/cardiomyopathy associated with mutations in plakophilin-2. Circulation 113(13):1641–1649

den Haan AD, Tan BY, Zikusoka MN, Llado LI, Jain R, Daly A, Tichnell C, James C, Amat-Alarcon N, Abraham T, Russell SD, Bluemke DA, Calkins H, Dalal D, Judge DP (2009) Comprehensive desmosome mutation analysis in north Americans with arrhythmogenic right ventricular dysplasia/cardiomyopathy. Circ Cardiovasc Genet 2(5):428–435

Djouadi F, Lecarpentier Y, Hebert JL, Charron P, Bastin J, Coirault C (2009) A potential link between peroxisome proliferator-activated receptor signalling and the pathogenesis of arrhythmogenic right ventricular cardiomyopathy. Cardiovasc Res 84(1):83–90

Ferrick DA, Neilson A, Beeson C (2008) Advances in measuring cellular bioenergetics using extracellular flux. Drug Discov Today 13(5–6):268–274

Garcia-Gras E, Lombardi R, Giocondo MJ, Willerson JT, Schneider MD, Khoury DS, Marian AJ (2006) Suppression of canonical Wnt/beta-catenin signaling by nuclear plakoglobin recapitulates phenotype of arrhythmogenic right ventricular cardiomyopathy. J Clin Invest 116(7):2012–2021

Go AS, Mozaffarian D, Roger VL, Benjamin EJ, Berry JD, Blaha MJ, Dai S, Ford ES, Fox CS, Franco S, Fullerton HJ, Gillespie C, Hailpern SM, Heit JA, Howard VJ, Huffman MD, Judd SE, Kissela BM, Kittner SJ, Lackland DT, Lichtman JH, Lisabeth LD, Mackey RH, Magid DJ, Marcus GM, Marelli A, Matchar DB, McGuire DK, Mohler ER 3rd, Moy CS, Mussolino ME, Neumar RW, Nichol G, Pandey DK, Paynter NP, Reeves MJ, Sorlie PD, Stein J, Towfighi A, Turan TN, Virani SS, Wong ND, Woo D, Turner MB (2014) Heart disease and stroke statistics-2014 update: a report from the American Heart Association. Circulation 129(3):e28–e292

Gregoire FM, Smas CM, Sul HS (1998) Understanding adipocyte differentiation. Physiol Rev 78(3):783–809

Itzhaki I, Maizels L, Huber I, Zwi-Dantsis L, Caspi O, Winterstern A, Feldman O, Gepstein A, Arbel G, Hammerman H, Boulos M, Gepstein L (2011) Modelling the long QT syndrome with induced pluripotent stem cells. Nature 471(7337):225–229

Kim C, Majdi M, Xia P, Wei KA, Talantova M, Spiering S, Nelson B, Mercola M, Chen HS (2010) Non-cardiomyocytes influence the electrophysiological maturation of human embryonic stem cell-derived cardiomyocytes during differentiation. Stem Cells Dev 19(6):783–795

Kim C, Wong J, Wen J, Wang S, Wang C, Spiering S, Kan NG, Forcales S, Puri PL, Leone TC, Marine JE, Calkins H, Kelly DP, Judge DP, Chen HS (2013) Studying arrhythmogenic right ventricular dysplasia with patient-specific iPSCs. Nature 494(7435):105–110

Knollmann BC (2013) Induced pluripotent stem cell-derived cardiomyocytes: boutique science or valuable arrhythmia model? Circ Res 112(6):969–976; discussion 976

Li D, Liu Y, Maruyama M, Zhu W, Chen H, Zhang W, Reuter S, Lin SF, Haneline LS, Field LJ, Chen PS, Shou W (2011) Restrictive loss of plakoglobin in cardiomyocytes leads to arrhythmogenic cardiomyopathy. Hum Mol Genet 20(23):4582–4596

Lombardi R, Dong J, Rodriguez G, Bell A, Leung TK, Schwartz RJ, Willerson JT, Brugada R, Marian AJ (2009) Genetic fate mapping identifies second heart field progenitor cells as a source of adipocytes in arrhythmogenic right ventricular cardiomyopathy. Circ Res 104(9): 1076–1084

Lopaschuk GD, Ussher JR, Folmes CD, Jaswal JS, Stanley WC (2010) Myocardial fatty acid metabolism in health and disease. Physiol Rev 90(1):207–258

Macfarlane DP, Forbes S, Walker BR (2008) Glucocorticoids and fatty acid metabolism in humans: fuelling fat redistribution in the metabolic syndrome. J Endocrinol 197(2):189–204

Marcus FI, Fontaine GH, Guiraudon G, Frank R, Laurenceau JL, Malergue C, Grosgogeat Y (1982) Right ventricular dysplasia: a report of 24 adult cases. Circulation 65(2):384–398

Marcus FI, McKenna WJ, Sherrill D, Basso C, Bauce B, Bluemke DA, Calkins H, Corrado D, Cox MG, Daubert JP, Fontaine G, Gear K, Hauer R, Nava A, Picard MH, Protonotarios N, Saffitz JE, Sanborn DM, Steinberg JS, Tandri H, Thiene G, Towbin JA, Tsatsopoulou A, Wichter T, Zareba W (2010) Diagnosis of arrhythmogenic right ventricular cardiomyopathy/ dysplasia: proposed modification of the Task Force Criteria. Eur Heart J 31(7):806–814

Marfella R, Portoghese M, Ferraraccio F, Siniscalchi M, Babieri M, Di Filippo C, D'Amico M, Rossi F, Paolisso G (2009) Thiazolidinediones may contribute to the intramyocardial lipid accumulation in diabetic myocardium: effects on cardiac function. Heart (Br Card Soc) 95(12): 1020–1022

Mercola M, Colas A, Willems E (2013) Induced pluripotent stem cells in cardiovascular drug discovery. Circ Res 112(3):534–548

Moretti A, Bellin M, Welling A, Jung CB, Lam JT, Bott-Flugel L, Dorn T, Goedel A, Hohnke C, Hofmann F, Seyfarth M, Sinnecker D, Schomig A, Laugwitz KL (2010) Patient-specific induced pluripotent stem-cell models for long-QT syndrome. N Engl J Med 363(15): 1397–1409

Moretti A, Laugwitz KL, Dorn T, Sinnecker D, Mummery C (2013) Pluripotent stem cell models of human heart disease. Cold Spring Harb Perspect Med 3(11):1–20

Murry CE, Reinecke H, Pabon LM (2006) Regeneration gaps: observations on stem cells and cardiac repair. J Am Coll Cardiol 47(9):1777–1785

Neubauer S (2007) The failing heart – an engine out of fuel. N Engl J Med 356(11):1140–1151

Okano H, Nakamura M, Yoshida K, Okada Y, Tsuji O, Nori S, Ikeda E, Yamanaka S, Miura K (2013) Steps toward safe cell therapy using induced pluripotent stem cells. Circ Res 112(3): 523–533

Onay-Besikci A (2006) Regulation of cardiac energy metabolism in newborn. Mol Cell Biochem 287(1–2):1–11

Park IH, Arora N, Huo H, Maherali N, Ahfeldt T, Shimamura A, Lensch MW, Cowan C, Hochedlinger K, Daley GQ (2008) Disease-specific induced pluripotent stem cells. Cell 134(5):877–886

Pettinelli P, Videla LA (2011) Up-regulation of PPAR-gamma mRNA expression in the liver of obese patients: an additional reinforcing lipogenic mechanism to SREBP-1c induction. J Clin Endocrinol Metab 96(5):1424–1430

Qyang Y, Martin-Puig S, Chiravuri M, Chen S, Xu H, Bu L, Jiang X, Lin L, Granger A, Moretti A, Caron L, Wu X, Clarke J, Taketo MM, Laugwitz KL, Moon RT, Gruber P, Evans SM, Ding S, Chien KR (2007) The renewal and differentiation of Isl1+ cardiovascular progenitors are controlled by a Wnt/beta-catenin pathway. Cell Stem Cell 1(2):165–179

Razani B, Zhang H, Schulze PC, Schilling JD, Verbsky J, Lodhi IJ, Topkara VK, Feng C, Coleman T, Kovacs A, Kelly DP, Saffitz JE, Dorn GW 2nd, Nichols CG, Semenkovich CF (2011) Fatty acid synthase modulates homeostatic responses to myocardial stress. J Biol Chem 286(35):30949–30961

Santostefano KE, Hamazaki T, Biel NM, Jin S, Umezawa A, Terada N (2015) A practical guide to induced pluripotent stem cell research using patient samples. Lab Invest 95(1):4–13

Sawant AC, Calkins H (2015) Sports in patients with arrhythmogenic right ventricular dysplasia/ cardiomyopathy and desmosomal mutations. Herz 40(3):402–409

Shiba Y, Fernandes S, Zhu WZ, Filice D, Muskheli V, Kim J, Palpant NJ, Gantz J, Moyes KW, Reinecke H, Van Biber B, Dardas T, Mignone JL, Izawa A, Hanna R, Viswanathan M, Gold JD, Kotlikoff MI, Sarvazyan N, Kay MW, Murry CE, Laflamme MA (2012) Human ES-cell-derived cardiomyocytes electrically couple and suppress arrhythmias in injured hearts. Nature 489(7415):322–325

Shinnawi R, Gepstein L (2014) iPCS cell modeling of inherited cardiac arrhythmias. Curr Treat Options Cardiovasc Med 16(9):331

Son NH, Park TS, Yamashita H, Yokoyama M, Huggins LA, Okajima K, Homma S, Szabolcs MJ, Huang LS, Goldberg IJ (2007) Cardiomyocyte expression of PPARgamma leads to cardiac dysfunction in mice. J Clin Invest 117(10):2791–2801

Souders CA, Bowers SL, Baudino TA (2009) Cardiac fibroblast: the renaissance cell. Circ Res 105(12):1164–1176

Swope D, Cheng L, Gao E, Li J, Radice GL (2012) Loss of cadherin-binding proteins beta-catenin and plakoglobin in the heart leads to gap junction remodeling and arrhythmogenesis. Mol Cell Biol 32(6):1056–1067

Takahashi K, Tanabe K, Ohnuki M, Narita M, Ichisaka T, Tomoda K, Yamanaka S (2007) Induction of pluripotent stem cells from adult human fibroblasts by defined factors. Cell 131(5):861–872

Veerman CC, Kosmidis G, Mummery CL, Casini S, Verkerk AO, Bellin M (2015) Immaturity of human stem-cell-derived cardiomyocytes in culture: fatal flaw or soluble problem? Stem Cells Dev 24(9):1035–1052

Waku T, Shiraki T, Oyama T, Maebara K, Nakamori R, Morikawa K (2010) The nuclear receptor PPARgamma individually responds to serotonin- and fatty acid-metabolites. EMBO J 29(19): 3395–3407

Willson TM, Lambert MH, Kliewer SA (2001) Peroxisome proliferator-activated receptor gamma and metabolic disease. Annu Rev Biochem 70:341–367

Yang X, Pabon L, Murry CE (2014) Engineering adolescence: maturation of human pluripotent stem cell-derived cardiomyocytes. Circ Res 114(3):511–523

Yazawa M, Hsueh B, Jia X, Pasca AM, Bernstein JA, Hallmayer J, Dolmetsch RE (2011) Using induced pluripotent stem cells to investigate cardiac phenotypes in Timothy syndrome. Nature 471(7337):230–234

Yu J, Vodyanik MA, Smuga-Otto K, Antosiewicz-Bourget J, Frane JL, Tian S, Nie J, Jonsdottir GA, Ruotti V, Stewart R, Slukvin II, Thomson JA (2007) Induced pluripotent stem cell lines derived from human somatic cells. Science (New York, N Y) 318(5858):1917–1920

Zhang Z, Stroud MJ, Zhang J, Fang X, Ouyang K, Kimura K, Mu Y, Dalton ND, Gu Y, Bradford WH, Peterson KL, Cheng H, Zhou X, Chen J (2015) Normalization of Naxos plakoglobin levels restores cardiac function in mice. J Clin Invest 125(4):1708–1712

Cardiac Arrhythmia Modelling Using iPS Cells

4

Shinsuke Yuasa

Abstract

Innovation in stem cell biology has greatly impacted on medical researches. The generation of induced pluripotent stem cell (iPSC) established the novel concept not only in regenerative medicine but also disease modelling. There are several reports that disease-specific iPSC-derived cells have recapitulated cellular phenotypes of wide varieties of diseases. The disease-specific iPSCs are utilised to understand unknown pathogenesis by in vitro examination of disease-specific iPSC-derived cells. It is also shown that disease-specific iPSC-derived cells can be applied to high throughput screening for new drugs. Early reports about disease modelling using iPSCs focused on various types of cardiac arrhythmia because it is difficult to model human arrhythmia in animal models such as a genetically engineered mouse and it is relatively clear to show the phenotypes of cardiac arrhythmia in iPSC-derived cardiomyocyte such as abnormal electrophysiological activities. In this chapter, we focus on the conceptual and practical issues about cardiac arrhythmia modelling using disease-specific iPSCs and discuss the future directions.

Keywords

Heart • Arrhythmia • iPS cell • Disease model

S. Yuasa, M.D., Ph.D. (✉)
Department of Cardiology, Keio University School of Medicine, 35-Shinanomachi, Shinjuku-ku, Tokyo 160-8582, Japan
e-mail: yuasa@keio.jp

K. Fukuda (ed.), *Human iPS Cells in Disease Modelling*,
DOI 10.1007/978-4-431-55966-5_4

4.1 Introduction

Induced pluripotent stem cells (iPSCs) are defined as pluripotent stem cells which can be generated from somatic cells by introducing reprogramming factors such as *OCT4*, *SOX2*, *KLF4*, and *MYC* (Takahashi and Yamanaka 2006). iPSCs definitely have the two essential characteristics of stem cells: infinite self-renewal capability and pluripotency. iPSCs can give rise to all cell types of three germ layers in vitro and in vivo, like embryonic stem cells (ESCs). iPSCs have definite identical genetic information with original somatic cells. Patient-specific iPSCs have advantages in regenerative medicine because patient-specific iPSC-derived cells showed immunocompatibility with host patients (Yuasa and Fukuda 2008). Reprogramming of somatic cells from patients with genetic mutations into iPSCs potentially enables us to generate disease-specific iPSCs and model the disease in vitro (Takahashi et al. 2007). Disease-specific iPSCs have opened new possibilities for continuous supplies of human diseased cells for disease models and drug screening. We can directly and repetitively analyse diseased cells using disease-specific iPSC-derived cells. In the past few years, many studies about human genetic diseases including cardiovascular diseases have yielded disease models using patient-specific iPSCs (Tiscornia et al. 2011; Egashira et al. 2012; Tanaka et al. 2014; Kodaira et al. 2015; Minegishi et al. 2013; Okata et al. 2013; Shimamoto et al. 2014; Iejima et al. 2015; Egashira et al. 2013). To date, many groups have reported that cellular phenotypes of genetic diseases can be recapitulated on patient-specific iPSC-derived cells in vitro and some papers also showed drug screenings using iPSCs for novel drug candidates.

To keep homeostasis in human, cardiomyocyte rhythmical contraction is indispensable. The cardiac rhythm abnormality can be a direct cause of human death. Rhythmical cardiomyocyte contraction is regulated by cardiomyocyte-specific electrical activity such as repetitive depolarization and repolarization. Many hereditary arrhythmias are caused by single mutations in ion channel genes. To understand the pathogenic mechanisms of cardiac arrhythmias, the experimental model is required for the analysis. In medical research, the animal model is widely used for the disease analysis such as the genetically engineered mouse. But there are many differences between human heart and mouse heart such as size, beating rate, the shape of action potential and ion channel expression. Therefore it is hardly model human arrhythmia in experimental animals. Patient-specific iPSC-derived cardiomyocyte is a human cardiomyocyte with patient genetic information, which would expect us to be able to model human arrhythmia in a dish. There are many attempts to model human heritable arrhythmia using patient-specific iPSCs and decipher the pathogenic mechanism. In this chapter, we review the published works related to disease modelling using patient-specific iPSCs in arrhythmias.

4.2 Disease Modelling by iPSCs

Because patient-specific iPSCs have identical genetic information with the host patients, any type of disease phenotypes can be theoretically recapitulated in genetically heritable disease. It is impossible to use patient heart samples endlessly for basic researches but disease-specific iPSCs can provide disease-specific cardiomyocytes infinitely. Therefore we highly expected the usage of iPSCs to model human cardiac disease. The disease modelling by using disease-specific iPSCs is in the predawn of medical research area and there are some requirements for a proof of concept study. It is preferred that the disease phenotypes can be easily and simply shown in cell culture dish and the analysis methods of diseased phenotypes have been established. Actually, most studies about disease modelling by using patient-specific iPSCs reported so far satisfied such requirements. Cardiac arrhythmias which were often selected for the study of disease modelling are inherited life-threatening diseases caused by functional impairment in cardiac ion channel and often cause sudden cardiac death by ventricular tachyarrhythmia.

But some disease phenotypes are hardly recapitulated because of several reasons. Firstly, in order to appropriately model the diseases, we inevitably differentiate disease-specific iPSCs to specific cell types which would show the disease phenotypes. But some specific cell types are hardly differentiated from iPSCs and the purification method of specific cell types is not established in many cell types. Secondly, the iPSC-derived cells are not fully matured and typically do not have adult phenotypes. It is easy to reproduce the phenotype of young diseases but difficult to reproduce the phenotype of adult disease because of immaturity of iPSC-derived cells. Thirdly, the disease phenotype which is mainly caused by a single gene mutation is suitable for the analysis but disease phenotype which is caused by environmental factor and epigenetic status is not suitable for modelling using iPSCs. It is difficult to reproduce the physiological environment around iPSC-derived cells. Epigenetic information of iPSC-derived cells would be partly renewed during the process of reprogramming and parental epigenetic information is hardly reproduced in iPSC-derived cells. Fourthly, the cellular interaction among different cell types is required to show the some disease phenotypes. It is technically possible to prepare some cell types but it is still difficult to prepare the many cell types. Lastly, to show the same disease phenotypes, the organ structure is required. It is still difficult to form a 3D structure by iPSC-derived cells.

4.3 Long QT Syndrome Modelling by iPS Cells

On the premise of cardiovascular disease modelling using patient-specific iPSCs, it is necessary to clarify that characteristics of human iPSC-derived cardiomyocytes are physiologically similar to human cardiomyocytes. ESCs are well-established pluripotent stem cells with a long history. ESC-derived cardiomyocytes is also well characterised and proved to be physiologically analogous to human cardiomyocytes, although not fully matured. Previous molecular and physiological studies have revealed that iPSC-derived cardiomyocytes have normal cardiomyocyte functional properties as in the case of ESC-derived cardiomyocytes (Seki et al. 2014). iPSC-derived cardiomyocytes have striated muscle structure identical to normal functional cardiomyocytes and expressed cardiac-specific proteins. iPSC-derived cardiomyocytes show three different action potential patterns such as atrial, nodal, and ventricular cells. Moreover contraction of iPSC-derived cardiomyocytes is regulated by physiological intracellular signalling including excitation-contraction coupling (EC-coupling) and those cardiomyocytes properly express typical ion channels. iPSC-derived cardiomyocytes respond to several drugs, which suggests that iPSC-derived cardiomyocytes can be used for in vitro pharmacologic testing (Tanaka et al. 2009). All these findings can indicate the validity of the study to analyse cardiovascular disease using patient-specific iPSC-derived cardiomyocytes.

Long QT syndrome (LQTS) was firstly selected for modelling using iPSCs as a proof-of-concept study because most LQTSs are caused by single mutations in ion channel genes and the disease phenotypes in iPSC-derived cardiomyocytes can be examined by established electrophysiological assay. Firstly, Moretti et al. showed that patient-specific iPSC-derived cardiomyocytes could recapitulate the disease phenotype in LQTS (Moretti et al. 2010). The authors generated iPSCs from two patients with LQTS type1 (LQTS1), who had autosomal dominant inheritance of a G569A missense mutation in *KCNQ1* gene-encoding IKs channel, which is related to the pathogenesis of LQTS. Individual cardiomyocytes derived from LQTS1 patient-specific iPSC (LQTS1-iPSC)-derived cardiomyocytes showed prolonged action potentials compared with control iPSC-derived cardiomyocytes. LQTS1-iPSC-derived cardiomyocytes had an increased susceptibility to arrhythmic event by catecholamine stimulation, which is one of the most important clinical features in LQTS1. This was a first report which showed that iPSCs can be used for disease modelling in cardiovascular diseases. Thereafter there were many reports which showed LQTS disease modelling using patient-specific iPSCs, in which the selected patients had mutated channels which had been already characterised by conventional experimental methods. Those reports confirmed that iPSC-derived cardiomyocytes can recapitulate disease phenotypes (Itzhaki et al. 2011; Yazawa et al. 2011).

4.4 Long QT Syndrome Type 1 Modelling by iPS Cells

In clinics, many patients have unknown mutations which give us no specific information about their disease phenotypes. To address whether iPSC technology could be used to characterise the disease phenotype with a novel mutated gene, we selected LQTS patients without family history and previous disease characterisation (Egashira et al. 2012). The patient was a 13-year-old boy who survived from cardiac arrest due to ventricular fibrillation with prolonged QT interval in electrocardiogram. Genetic testing revealed that the patient had a novel heterozygous mutation in *KCNQ1*, 1893delC. Then iPSC was generated from the patient and patient-specific iPSCs were differentiated into cardiomyocytes. The electrophysiological function was measured by multielectrode array system and showed that duration of field potential (FPD) was markedly prolonged in LQTS patient-specific iPSC (LQTS-iPSC)-derived cardiomyocytes, as compared with control, which suggested that LQTS-iPSC-derived cardiomyocytes recapitulated the phenotype of LQTS. Next we tried to identify a responsible channel for disease phenotype by precise examination of several drug responses. IKs channel blocker prolonged FPD in control cardiomyocytes but didn't affect FPD in LQTS-iPSC-derived cardiomyocytes, which suggested that IKs was functionally impaired in LQTS-iPSC-derived cardiomyocytes. In general, IKr and IKs channel work in complementary fashion in repolarization of cardiomyocytes, which is known as repolarization reserve. The IKr blocker administration testing confirmed that IKr channels had normal function, which suggested that IKr would compensate for the repolarization of LQTS-iPSCs-derived cardiomyocytes. Adrenergic stimulation increased arrhythmogenic events of LQTS-iPSC-derived cardiomyocytes, which also suggested the recapitulation of patient phenotypes. These findings suggested that patient's IKs channel was functionally impaired in patient cardiomyocytes and diagnosis for the patient was LQTS type1. In our study, we demonstrated that iPSCs could be useful to characterise the electrophysiological cellular phenotype of patient with a novel mutation. We performed functional analysis of the novel mutation by using patient-specific iPSCs, which may support the clinical diagnosis with a novel mutation. Drug testing in arrhythmic patients has some risks to induce arrhythmic events. But this system allowed us to perform several drug administration tests on patient-specific iPSC-derived cardiomyocytes, which might substitute the drug tests for patients in future. And more patient-specific iPSC technology can be used for drug evaluation and monitoring as a personalised medicine.

4.5 Conclusions

Although iPSC technology is an attractive tool for analysing human genetic diseases, technological innovation remains an important requirement for the utilisation of iPSCs in routine medical practice. Patient-specific iPSC technology enables us to directly and repetitively approach to diseased cells and has great possibility to elucidate novel disease pathogenesis and develop new therapeutics. However, in terms of effort, cost, and time exhausted in current studies using iPSCs, it is still premature to realise the routine clinical usage. But iPSC technology is now undergoing steady improvements to overcome the hurdles. In the near future, we will open a gate of innovative medicine by using disease-specific iPSCs.

References

Egashira T, Yuasa S, Suzuki T, Aizawa Y, Yamakawa H, Matsuhashi T, Ohno Y, Tohyama S, Okata S, Seki T, Kuroda Y, Yae K, Hashimoto H, Tanaka T, Hattori F, Sato T, Miyoshi S, Takatsuki S, Murata M, Kurokawa J, Furukawa T, Makita N, Aiba T, Shimizu W, Horie M, Kamiya K, Kodama I, Ogawa S, Fukuda K (2012) Disease characterization using LQTS-specific induced pluripotent stem cells. Cardiovasc Res 95(4):419–429. doi:10.1093/cvr/cvs206

Egashira T, Yuasa S, Fukuda K (2013) Novel insights into disease modeling using induced pluripotent stem cells. Biol Pharm Bull 36(2):182–188. doi:10.1248/bpb.b12-00960

Iejima D, Itabashi T, Kawamura Y, Noda T, Yuasa S, Fukuda K, Oka C, Iwata T (2015) HTRA1 (High Temperature Requirement A Serine Peptidase 1) gene is transcriptionally regulated by insertion/deletion nucleotides located at the 3′ end of the ARMS2 (age-related maculopathy susceptibility 2) gene in patients with age-related macular degeneration. J Biol Chem 290(5): 2784–2797. doi:10.1074/jbc.M114.593384

Itzhaki I, Maizels L, Huber I, Zwi-Dantsis L, Caspi O, Winterstern A, Feldman O, Gepstein A, Arbel G, Hammerman H, Boulos M, Gepstein L (2011) Modelling the long QT syndrome with induced pluripotent stem cells. Nature 471(7337):225–229. doi:http://www.nature.com/nature/journal/v471/n7337/abs/10.1038-nature09747-unlocked.html#supplementary-information

Kodaira M, Hatakeyama H, Yuasa S, Seki T, Egashira T, Tohyama S, Kuroda Y, Tanaka A, Okata S, Hashimoto H, Kusumoto D, Kunitomi A, Takei M, Kashimura S, Suzuki T, Yozu G, Shimojima M, Motoda C, Hayashiji N, Saito Y, Goto Y-i, Fukuda K (2015) Impaired respiratory function in MELAS-induced pluripotent stem cells with high heteroplasmy levels. FEBS Open Biol 5(0):219–225. doi:10.1016/j.fob.2015.03.008

Minegishi Y, Iejima D, Kobayashi H, Chi Z-L, Kawase K, Yamamoto T, Seki T, Yuasa S, Fukuda K, Iwata T (2013) Enhanced optineurin E50K–TBK1 interaction evokes protein insolubility and initiates familial primary open-angle glaucoma. Hum Mol Genet 22(17): 3559–3567. doi:10.1093/hmg/ddt210

Moretti A, Bellin M, Welling A, Jung CB, Lam JT, Bott-Flügel L, Dorn T, Goedel A, Höhnke C, Hofmann F, Seyfarth M, Sinnecker D, Schömig A, Laugwitz K-L (2010) Patient-specific induced pluripotent stem-cell models for long-QT syndrome. N Engl J Med 363(15): 1397–1409. doi:10.1056/NEJMoa0908679

Okata S, Yuasa S, Yamane T, Furukawa T, Fukuda K (2013) The generation of induced pluripotent stem cells from a patient with KCNH2 G603D, without LQT2 disease associated symptom. J Med Dent Sci 60(1):17–22

Seki T, Yuasa S, Kusumoto D, Kunitomi A, Saito Y, Tohyama S, Yae K, Kishino Y, Okada M, Hashimoto H, Takei M, Egashira T, Kodaira M, Kuroda Y, Tanaka A, Okata S, Suzuki T, Murata M, Fujita J, Fukuda K (2014) Generation and characterization of functional cardiomyocytes derived from human T cell-derived induced pluripotent stem cells. PLoS ONE 9(1):e85645. doi:10.1371/journal.pone.0085645

Shimamoto A, Kagawa H, Zensho K, Sera Y, Kazuki Y, Osaki M, Oshimura M, Ishigaki Y, Hamasaki K, Kodama Y, Yuasa S, Fukuda K, Hirashima K, Seimiya H, Koyama H, Shimizu T, Takemoto M, Yokote K, Goto M, Tahara H (2014) Reprogramming suppresses premature senescence phenotypes of werner syndrome cells and maintains chromosomal stability over long-term culture. PLoS ONE 9(11):e112900. doi:10.1371/journal.pone.0112900

Takahashi K, Yamanaka S (2006) Induction of pluripotent stem cells from mouse embryonic and adult fibroblast cultures by defined factors. Cell 126(4):663–676. doi:10.1016/j.cell.2006.07.024

Takahashi K, Tanabe K, Ohnuki M, Narita M, Ichisaka T, Tomoda K, Yamanaka S (2007) Induction of pluripotent stem cells from adult human fibroblasts by defined factors. Cell 131(5):861–872. doi:10.1016/j.cell.2007.11.019

Tanaka T, Tohyama S, Murata M, Nomura F, Kaneko T, Chen H, Hattori F, Egashira T, Seki T, Ohno Y, Koshimizu U, Yuasa S, Ogawa S, Yamanaka S, Yasuda K, Fukuda K (2009) In vitro pharmacologic testing using human induced pluripotent stem cell-derived cardiomyocytes. Biochem Biophys Res Commun 385(4):497–502. doi:10.1016/j.bbrc.2009.05.073

Tanaka A, Yuasa S, Mearini G, Egashira T, Seki T, Kodaira M, Kusumoto D, Kuroda Y, Okata S, Suzuki T, Inohara T, Arimura T, Makino S, Kimura K, Kimura A, Furukawa T, Carrier L, Node K, Fukuda K (2014) Endothelin-1 induces myofibrillar disarray and contractile vector variability in hypertrophic cardiomyopathy–induced pluripotent stem cell–derived cardiomyocytes. J Am Heart Assoc 3(6), e001263. doi:10.1161/jaha.114.001263

Tiscornia G, Vivas EL, Belmonte JCI (2011) Diseases in a dish: modeling human genetic disorders using induced pluripotent cells. Nat Med 17(12):1570–1576

Yazawa M, Hsueh B, Jia X, Pasca AM, Bernstein JA, Hallmayer J, Dolmetsch RE (2011) Using induced pluripotent stem cells to investigate cardiac phenotypes in Timothy syndrome. Nature 471(7337):230–234. doi:http://www.nature.com/nature/journal/v471/n7337/abs/10.1038-nature09855-unlocked.html#supplementary-information

Yuasa S, Fukuda K (2008) Recent advances in cardiovascular regenerative medicine: the induced pluripotent stem cell era. Expert Rev Cardiovasc Ther 6(6):803–810. doi:10.1586/14779072.6.6.803

iPSC Disease Modeling of Laminopathies

5

Guang-Hui Liu, Jiping Yang, Zhichao Ding, Alejandro Ocampo, Jing Qu, Keiichiro Suzuki, and Juan Carlos Izpisua Belmonte

Abstract

Nuclear lamins, the major components of the nuclear lamina, participate in maintaining the nuclear structure and are involved in essential cellular processes, including DNA replication and chromatin organization. To date, more than 15 human diseases classified as laminopathies have been linked to mutations in components of the nuclear lamina. With the development of induced pluripotent stem cell (iPSC) and targeted gene editing technologies, novel disease

---

Guang-Hui Liu and Jingping Yang contributed equally with all other contributors.

G.-H. Liu (✉) • Z. Ding
National Laboratory of Biomacromolecules, Institute of Biophysics, Chinese Academy of Sciences, Beijing 100101, China
e-mail: ghliu@ibp.ac.cn

J. Yang
National Laboratory of Biomacromolecules, Institute of Biophysics, Chinese Academy of Sciences, Beijing 100101, China

State Key Laboratory of Stem Cell and Reproductive Biology, Institute of Zoology, Chinese Academy of Sciences, Beijing 100101, China

A. Ocampo • J.C.I. Belmonte (✉)
Gene Expression Laboratory, Salk Institute for Biological Studies, 10010 N. Torrey Pines Rd, La Jolla, CA 92037, USA
e-mail: belmonte@salk.edu

J. Qu
State Key Laboratory of Stem Cell and Reproductive Biology, Institute of Zoology, Chinese Academy of Sciences, Beijing 100101, China

K. Suzuki
Gene Expression Laboratory, Salk Institute for Biological Studies, 10010 N. Torrey Pines Rd, La Jolla, CA 92037, USA

Universidad Católica San Antonio de Murcia (UCAM), Campus de los Jerónimos, N° 135, Guadalupe 30107, Murcia, Spain

K. Fukuda (ed.), *Human iPS Cells in Disease Modelling*,
DOI 10.1007/978-4-431-55966-5_5

models of laminopathies have been generated. These models will improve our understanding of the pathological mechanism of laminopathies and may eventually lead to the discovery of new therapeutic approaches for the treatment of these devastating diseases.

Keywords
iPSC • Disease model • Laminopathy

5.1 Introduction

Cellular organelles have emerged as part of the evolution of eukaryotic cells, which enable complex specialized cellular functions to be performed in an accurate manner. The nucleus where the cell genetic material is located is separated from cytoplasm by the nuclear envelope. Beneath the nuclear envelope, there is a fibrillar structural network known as the nuclear lamina. The nuclear lamina is mainly composed of nuclear lamins, which are a type V intermediate filament protein (Butin-Israeli et al. 2012; Gomez-Cavazos and Hetzer 2012; Schreiber and Kennedy 2013). The nuclear lamina not only participates in maintaining the nuclear structure but is also involved in various cellular processes including DNA replication and chromatin organization (Gomez-Cavazos and Hetzer 2012; Schreiber and Kennedy 2013). Due to its fundamental role as part of the nuclear envelope, so far, more than 15 human diseases have been linked to the nuclear lamina. These diseases result from mutations in the components of the nuclear lamina and are classified as laminopathies (Butin-Israeli et al. 2012).

Disease modeling of laminopathies may serve to improve our understanding of the molecular mechanism of these disorders and be used as a platform for the discovery of novel treatments. Previous disease models such as animal models have inevitable and intrinsic disadvantages that limit their further application. On the other hand, recent studies have demonstrated the potential use of induced pluripotent stem cells (iPSCs) derived from autologous somatic cells, followed by cellular differentiation, as a new strategy to model diseases (Ho et al. 2011; Liu et al. 2011a; Nissan et al. 2012; Siu et al. 2012; Xiong et al. 2013; Zhang et al. 2011). Moreover, with the improvement of targeted gene-editing technologies, it is now possible for the generation of "disease cells" by knocking in disease-causing mutations in healthy cells (Liu et al. 2011b, 2012; Soldner et al. 2011). It is clear that using iPSC to model diseases represents a valuable tool to study disease pathogenesis and perform drug screens in the search for novel treatments and clinical therapies.

5.2 Nuclear Lamins

Nuclear lamins are the major components of the nuclear lamina and interact with heterochromatin and the inner nuclear membrane (Butin-Israeli et al. 2012). While most lamins are assembled at the nuclear lamina, some lamins are located throughout nucleoplasm. Nuclear lamins can be divided into A- and B-type lamins. A-type lamins include lamin A and lamin C which are alternative splicing variants of the *LMNA* gene, while B-type lamins contain lamin B1 and lamin B2 which are encoded by *LMNB1* and *LMNB2* genes, respectively (Dechat et al. 2010).

Lamins undergo a process of maturation including several posttranslational modifications to be functionally mature (Schreiber and Kennedy 2013; Young et al. 2006). The first step in this process is the farnesylation of the cysteine residue in the C-terminal CaaX motif. After farnesylation, the aaX of the CaaX motif is cleaved off by lamin-specific endoproteases (ZMPSTE24 cleaves pre-lamin A and RCE1 cleaves pre-lamin B), and subsequently, a new carboxymethyl group is transferred to the C-terminal cysteine residue. The maturation of lamin A ends with a second proteolytic event that removes another 15 C-terminal amino acids, which is mediated also by ZMPSTE24 (Young et al. 2006). Since lamin C is the alternative splicing variant of pre-lamin A and it lacks the C-terminus, no farnesylation and further modifications occur during lamin C maturation. However, B-type lamins avoid the second proteolysis and retain the farnesyl group that might help in the anchoring of the nuclear lamina to inner nuclear membrane (Jung et al. 2013). For this reason, it is suggested that the posttranslational modification of lamins is crucial for the assembly and organization of nuclear lamina as well as for the interactions between nuclear lamina and the nuclear envelope.

The lamin composition of the nuclear lamina differs among different cell types, although there seems to be no difference in the distribution of lamins. While A-type lamins are transcriptionally repressed in a pluripotent state, they are activated upon differentiation (Liu et al. 2011a; Niedernhofer et al. 2011; Zhang et al. 2011). On the contrary, B-type lamins are widely expressed in almost all cell lineages (Young et al. 2012). It is reported that the binding of lamin A to several important transcription factors regulates cellular processes like differentiation. Interestingly, hundreds of mutations in A-type lamins have been linked to serious human genetic disorders, highlighting the importance of A-type lamins in development (Butin-Israeli et al. 2012; Schreiber and Kennedy 2013). B-type lamins might be associated with some functions such as DNA replication and cell proliferation. Along this line, lamin B interacts with the nuclear envelope during mitosis, while lamin A/C does not (Gerace et al. 1978). In addition, lamin B might participate in the development of specific tissues such as the brain and lung. Lamin B1- or lamin B2-deficient mice show defects in brain development with neuronal migration severely impaired (Coffinier et al. 2010, 2011; Jung et al. 2013; Young et al. 2012). Lamin B1 null mice and the non-farnesylated lamin B1 mice also develop defective lungs with fewer alveoli (Jung et al. 2013; Vergnes et al. 2004). Interestingly, honeycomb distribution of lamin B1 is observed in the brain tissues of non-farnesylated lamin B1 mice (Jung et al. 2013).

5.3 Laminopathy

Laminopathies are a group of rare, severe human genetic disorders caused by mutations in genes encoding for protein components of the nuclear lamina including nuclear lamins and their interacting proteins such as emerin and nesprin. So far, more than 15 laminopathies have been discovered ranging from muscle diseases such as Emery-Dreifuss Muscular Dystrophy (EDMD) to premature aging disorders such as Hutchinson-Gilford Progeria Syndrome (HGPS) (Butin-Israeli et al. 2012; Schreiber and Kennedy 2013; Young et al. 2006).

Until now, over 200 disease-causing mutations in the *LMNA* gene have been reported to induce these human diseases. Most mutation-caused diseases are tissue specific, affecting muscle, adipose, heart, and neurons, respectively. Some mutations in the *LMNA* gene can influence the growth and development of individuals, resulting in premature aging disorders such as HGPS and atypical Werner syndrome (AWS) (Butin-Israeli et al. 2012). Patients suffering from these diseases present complex symptoms including atherosclerosis, wrinkled skin, hearing loss, and cardiovascular problems (Burtner and Kennedy 2010; Merideth et al. 2008). Of note, HGPS is the most common laminopathy associated with mutations in A-type lamin. To date, a number of useful human HGPS disease models have been generated with the help of iPSC technology (Ho et al. 2011; Liu et al. 2011a; Zhang et al. 2011).

Unlike lamin A, the B-type lamins are rarely related to diseases. Considering lamin B null mice die after birth, it is possible that B-type lamins are indispensable for individual viability (Butin-Israeli et al. 2012; Vergnes et al. 2004). However, duplication of the *LMNB1* gene results in autosomal dominant adult-onset demyelinating leukodystrophy (ADLD) characterized by defects in the central nervous system (CNS) (Padiath et al. 2006), where overexpressed lamin B1 triggers the degradation of myelin and disrupts the normal function of CNS. In addition, mutations in the *LMNB2* gene have been associated with lipodystrophy (Hegele et al. 2006).

Mutations of nuclear lamin interacting proteins have been linked to laminopathies as well. For instance, the striated muscle disease EDMD is related to mutations in *EMD* which encodes emerin, *SYNE1* which encodes nesprin-1, and *SYNE2* which encodes nesprin-2 (Nagano et al. 1996; Zhang et al. 2007). Because of their interaction with nuclear lamina, it is likely that mutant proteins indirectly affect lamin's function with pathological consequences.

5.4 Diseases Modeling of Laminopathies

iPSC technologies possess an incredible potential to model diseases using autologous patient-derived somatic cells in culture. By ectopic expression of several reprogramming factors and/or with small chemicals, somatic cells can be reprogrammed into a pluripotent state with self-renewal capability and the potential to differentiate into diverse cell lineages (Yamanaka and Blau 2010). Since many diseases are tissue-specific and due to the difficulty of gaining access to patient-specific samples, iPSC models may represent an unlimited source of "disease cells" and facilitate mechanistic studies as well as drug screening.

5.4.1 Hutchinson-Gilford Progeria Syndrome (HGPS)

Hutchinson-Gilford Progeria Syndrome (HGPS) is a severe human premature aging disease caused by a mutation in the *LMNA* gene (Burtner and Kennedy 2010; Kudlow et al. 2007; Merideth et al. 2008). HGPS symptoms resemble those of aging including hair loss, wrinkled skin, atherosclerosis, and cardiovascular problems but are manifested at a very early age. HGPS results from a single point mutation that leads to the interruption of posttranslational modifications essential for the maturation of lamin A. The most classical mutation causing HGPS is C1824T. Although this is a silent mutation (G608G), it generates a cryptic splicing site in exon 11 of *LMNA* which results in the production of a truncated lamin A called progerin (Young et al. 2006). Progerin, also known as the premature aging protein, is farnesylated and lacks 50 amino acids in the C-terminus compared with mature lamin A. Interestingly, progerin accumulates not only in HGPS patients but also in normal individuals during physiological aging (Olive et al. 2010; Scaffidi and Misteli 2006). Although it is believed that progerin acts through a gain-of-function toxicity to the cells, the molecular mechanism remains to be elucidated. At present, several hypotheses have been postulated. One of these hypotheses focus on the nuclear structure proposing that expression of progerin transforms the peripheral meshwork of the nucleus and alters the organization of heterochromatin that normally interacts with nuclear lamina (Dechat et al. 2008). As a consequence, the nuclear genome might become unstable and gene expression might be misregulated, affecting cellular proliferation and DNA repair, leading to senescence. Another hypothesis suggests that progerin might function as a transcription factor or cofactor and somehow regulate its downstream factors (Andres and Gonzalez 2009). As an example, progerin specifically downregulates the late adipogenic regulators leading to the loss of adipose tissue in HGPS patients (Xiong et al. 2013). Based on the potential availability through the use of iPSC technologies, we are now gaining more insight about the molecular mechanism of HGPS.

Since HGPS is a fatal laminopathy that presents common symptoms to other laminopathies such as striated muscle diseases and lipodystrophy as well as physiological aging, its study might provide new clues about not only premature aging but also normal aging. For this reason, we and other groups have generated iPSC models of HGPS to help us better understand, diagnose, and treat these terrible diseases.

HGPS patients present symptoms affecting specific tissues, including muscle, adipose tissue, and heart, which belong to mesenchymal lineages (Merideth et al. 2008). However, HGPS patients show mild or no symptoms in other tissues such as nerves. Through the use of reprogrammed patient cells and subsequent differentiation to mesenchymal stem cells or mesenchymal lineage-specific cells, it is now possible to unveil the pathogenic mysteries that determine affected cell types in HGPS. So far, recent studies have done some work with the following cell lines.

5.4.1.1 Fibroblast-Derived iPSCs

HGPS fibroblasts express high levels of progerin and show abnormal nuclear morphology (Liu et al. 2011a; Scaffidi and Misteli 2005; Zhang et al. 2011). Compared to normal fibroblasts, "blebbing" on the nuclear envelope is observed and the expressions of the nuclear lamina components lamin B1 and Lap2β are repressed (Liu et al. 2011a). Meanwhile, heterochromatin markers H3K9me3, HP1a, and HDAC1 and proliferation marker Ki67 are greatly reduced.

Recently, we and other groups have generated iPSC lines from fibroblasts of HGPS patients by viral transduction of the four Yamanaka factors – Oct4, Sox2, Klf4, and cMyc (Ho et al. 2011; Liu et al. 2011a; Zhang et al. 2011). Interestingly, there is a significant reduction in reprogramming efficiency between HGPS and normal fibroblasts where late-passage cells can hardly generate iPSCs (Zhang et al. 2011). Along this line, cell senescence seems to be an obstacle to properly reprogramming. Once reprogrammed into a pluripotent state, HGPS iPSC lines express pluripotency markers such as OCT4, SOX2, NANOG, and SEEA4 (Liu et al. 2011a; Zhang et al. 2011). In vitro embryoid body (EB) formation and in vivo teratoma formation

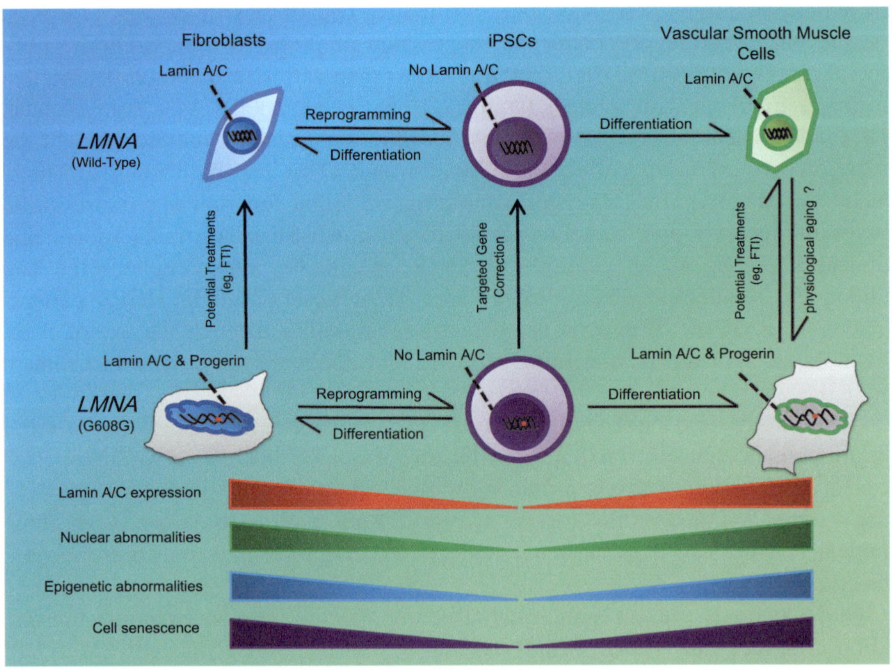

Fig. 5.1 Schematic demonstration of reprogramming and redifferentiation of HGPS fibroblasts

verified the pluripotency of iPSCs. These iPSC lines can be cultured in a culture dish for more than 50 passages without affecting their pluripotency.

LMNA is not expressed in human embryonic stem cells (hESCs); therefore, progerin, which is a processing product of *LMNA* expression, should not be detected in human pluripotent stem cells (Constantinescu et al. 2006). Accordingly, in HGPS iPSCs, progerin is silenced and abnormal nuclear morphologies are rescued despite the presence of the mutation in the *LMNA* gene (Liu et al. 2011a; Zhang et al. 2011). In addition, all other aspects such as epigenetic marks, proliferation, and nuclear lamina are similar to normal fibroblast-derived iPSCs. Moreover, genome-wide methylation analysis shows indistinguishable patterns between HGPS iPSCs and control iPSCs, while significant differences are seen between wild-type and HGPS fibroblasts (Liu et al. 2011a) (Fig. 5.1).

5.4.1.2 Vascular Smooth Muscle Cells

Atherosclerosis is a common symptom of aging-related cardiovascular disease (Gorenne et al. 2006). HGPS patients suffer from this medical condition as well. Since the degeneration of vascular smooth muscle cells (SMCs) is one of the characteristics of atherosclerosis, the disease model of HGPS-associated atherosclerosis focuses on the vascular smooth muscle cells. Serial passaging of SMCs is performed to model cellular senescence, progerin is accumulated in these SMCs, and the frequency of misshapen nuclei increases (Liu et al. 2011a). At later passages, both proliferation markers Ki67 and telomere length are reduced, coinciding with the appearance of typical characteristics of premature senescence. Moreover, when HGPS patient-derived SMCs are treated with different stress conditions, including hypoxia and hypoxia with substrate deprivation, cellular senescence is induced as shown by senescence-associated beta-galactosidase staining (Zhang et al. 2011). In addition, when SMCs are subjected to recurrent electrical stimulation to simulate in vivo stress, HGPS SMCs show abnormal nuclei and severe senescence compared to normal SMCs (Zhang et al. 2011). Specifically, DNA-dependent protein kinase catalytic subunit (DNAPKcs) is identified as a new downstream target of progerin, which provides new clues about the molecular mechanism behind HGPS (Liu et al. 2011a) (Fig. 5.1).

5.4.1.3 Adipocytes

HGPS patients present partial or total loss of adipose tissue and usually have thin and small stature. Adipocytes are the major component of adipose tissue which stores lipids and regulates energy homeostasis. The formation of adipocytes involves initial determination and terminal differentiation which are controlled by independent factors (Cristancho and Lazar 2011; Rosen and MacDougald 2006). C/EBPβ and C/EBPδ participate in the determination of adipocyte lineage, while the terminal differentiation step is dominated by PPARγ and C/EBPα. To model lipodystrophy in HGPS, HGPS iPSCs are induced into adipocyte lineages. Upon differentiation, expression of progerin and blebbing of the nucleus are restored as expected (Xiong et al. 2013). Although lipid droplets can be seen in HGPS cells, at later stages, there are fewer lipid droplets in HGPS cells compared to normal cells.

Additionally, binucleated cells and cellular senescence are observed in HGPS adipocytes. These imply failure of the terminal HGPS adipocyte differentiation. To understand the role of progerin in adipocyte differentiation, gene expressions of key regulators involved in this process are analyzed. While early regulators C/EBPβ and C/EBPδ are activated in both HGPS and normal cells in agreement with successful initiation of determination (Xiong et al. 2013), a later terminal differentiation step is hampered by downregulation of the core regulators PPARγ and C/EBPα in HGPS adipocytes. In line with the loss of fat during physiological aging of certain people, it is well known that age might affect the capacity of formation and the quality of adipocytes (Nakamura et al. 2012). In addition, old donor-derived adipocytes are capable of storing fewer lipids than young donor-derived adipocytes.

5.4.1.4 Neural Cells

HGPS patients rarely present neurodegenerative diseases, which often manifest during physiological aging (Burtner and Kennedy 2010; Merideth et al. 2008; Young et al. 2012). In agreement with this observation, when neural stem or progenitor cells are differentiated from HGPS iPSCs, no differences are observed between HGPS neural progenitors and control neural progenitors including the proliferation marker Ki67, neural marker nestin, as well as nuclear morphology (Nissan et al. 2012; Zhang et al. 2011). Moreover, further differentiation leads to cells expressing normal neuronal markers such as Tuj1. Importantly, both lamin A/C and progerin are downregulated in neural lineages and are completely absent in motor neurons. This evidence supports the idea that there exists some mechanism to suppress the expression of progerin in neural lineages so that no neurodegenerative symptoms are observed in HGPS patients. A potential explanation of the low expression of toxic progerin involves the presence of microRNA (Jung et al. 2012; Nissan et al. 2012). MicroRNAs are 20–25 nucleotide, noncoding RNAs which target sequence-specific mRNAs and guide translational repression or mRNA degradation (Rana 2007). When the 3′ untranslated region (UTR) sequences (also known as the seed sequences) of lamin A were input into prediction software to find potential microRNAs which could align the seed sequences, two candidate microRNAs, miR-129-5p and miR-9, were identified, and the latter has been found to play important roles in neural development (Jung et al. 2012; Young et al. 2012). MicroRNA expression analysis shows that miR-9 is expressed in HGPS neural stem cells and neurons, while absent in HGPS fibroblasts. More importantly, progerin and the number of blebbing nuclei are decreased upon ectopic expression of miR-9 in mesenchymal stem cells (Nissan et al. 2012). This study provides, to some extent, an answer to the question of why HGPS patients rarely suffer from neurodegenerative diseases and offers a potential strategy to treat HGPS through the use of microRNAs.

5.4.2 Other Laminopathies

In addition to HGPS, models of other laminopathies, such as dilated cardiomyopathy, have also been generated (Ho et al. 2011; Siu et al. 2012).

5.4.2.1 Dilated Cardiomyopathy (DCM)

DCM is a cardiac disease characterized by a weak and enlarged heart. Five to ten percentage of familial DCM cases are *LMNA*-associated DCM and are caused by several mutations in *LMNA* genes (Fatkin et al. 1999). To date, two DCM models bearing different mutations in *LMNA* have been generated (Ho et al. 2011; Siu et al. 2012). The first of these models is based on a novel GCCA insertion in base 50 that creates a premature stop codon. The haploinsufficiency of lamin A results in dilated ventricle and conduction system defects. So far, no further mechanistic study has been reported about DCM caused by this mutation (Ho et al. 2011). Another model is based on a nonsense mutation R225X which also leads to the haploinsufficiency of lamin A. Despite the presence of abnormal nuclei, beating cardiac cells can be obtained upon cardiac differentiation from iPSCs. Although under normal conditions, there is no major difference between control and DCM cardiomyocytes (Siu et al. 2012); when treated with electrical stimulation, a significantly increased senescence of DCM cardiomyocytes is observed. Additionally, electrical stimulation results in more severe cell apoptosis in the DCM context. Addition of MEK1/2 inhibitors, which block the ERK1/2 pathway, successfully rescues the accelerated apoptosis caused by electrical stimulation in DCM cardiomyocytes, supporting a notion that lack of lamin A could somehow influence the ERK1/2 pathway reducing the tolerance for electrical stress in cardiomyocytes (Siu et al. 2012).

5.4.2.2 Atypical Werner Syndrome (AWS)

The atypical form of Werner syndrome (WS) is known as AWS. Although the disease-causing genes are different between Werner syndrome and AWS (WS is caused by mutation in *WRN*, while AWS is the result of *LMNA* mutation), both of them are premature aging syndromes with similar features (Mounkes and Stewart 2004). Cellular phenotypes of AWS fibroblasts are common to HGPS cells including abnormal nuclear morphology and cellular senescence. Although normal nuclear morphology is present in a pluripotent state, upon redifferentiation into secondary fibroblasts, all abnormalities are recapitulated (Ho et al. 2011). Apart from this basic observation, to comprehensively understand the pathogenic mechanisms of syndromes such as AWS and HGPS that affect multiple tissues, global genomic, transcriptomic, and proteomic studies should be carried out.

5.4.3 Targeted Gene Editing and Disease Modeling

Despite the outstanding capabilities of iPSC technologies in disease modeling and their potential applications in scientific studies and clinical therapy, iPSC technologies have intrinsic disadvantages. For instance, most studies utilize iPSCs from healthy donors with similar age whose genetic background and genotypes are extremely different from those of patient-derived iPSCs. A possible solution to this problem could be based on the repair of "disease cells" by correcting the mutation responsible for the disease or alternatively to create "disease cells" by knocking in this mutation in healthy iPSCs.

Targeted correction or creation of "disease cells" requires the assistance of homologous recombination (HR) which has been successfully applied for the generation of animal models (Butin-Israeli et al. 2012). However, when applied to human cells, the efficiency of HR becomes extremely low (Goulburn et al. 2011; Ruby and Zheng 2009; Xue et al. 2009). To enhance the efficiency of HR, DNA double-stand breaks (DSBs) are induced in the genome with many gene-editing technologies (ZFN/TALEN/CRISPR) (Handel and Cathomen 2011; Hockemeyer et al. 2011; Mali et al. 2013; Zou et al. 2009), however, that might result in off-targeted breaks and lead to unexpected changes in the genome. Additionally, since there are more than 200 potential disease-causing mutations in a small region of *LMNA*, it will be desirable to develop special tools that would allow for the replacement of large genomic regions by HR.

Recently, a robust tool known as helper-dependent adenoviral vectors (HDAdVs) has been demonstrated to overcome most of the potential issues described above. HDAdVs have already been successfully used in human iPSCs and human mesenchymal stem cells (Liu et al. 2011b, 2012). HDAdV-mediated homologous recombination is DSB independent, which ensures the stability of the genome. Due to the complete removal of viral genes from the vector's genome, generated cell lines are absolutely integration-free (Suzuki et al. 2008). In addition, a long range of homology arms can be cloned into HDAdV vectors allowing for the correction of multiple mutations at the same time. Importantly, this approach shows tremendously high efficiency despite the usage of long homology arms. Using this strategy, we have successfully corrected the *LMNA* mutation in HGPS and AWS iPSCs. After correction, progerin expression in HGPS cells is not detected and no disease-related phenotypes are observed in the derivatives differentiated from corrected HGPS iPSCs, highlighting the power of HDAdV-mediated gene editing (Liu et al. 2011b). Moreover, HDAdV-mediated gene editing enables the creation of "disease cells" that are normally hard to access. Predesigned homology arms containing certain mutations could be used to knock in disease-causing mutations into normal human genome using HDAdV. This idea has been applied for the creation of Parkinson's disease cells containing a mutation in the leucine-rich repeat kinase 2 (LRRK2) (Liu et al. 2012). Disease phenotypes of Parkinson's disease iPSCs are recapitulated after targeted knock-in of the LRRK2 (G2019S) mutation in human embryonic stem cells. In this case, both "disease cells" and the normal cells share the same genetic background and therefore provide an unbiased system for mechanistic studies and drug screening. Otherwise, due to its safety and efficiency, in vitro cultured corrected neural progenitor cells might be applied for future cell transplantation as treatment for diseases (Pan et al. 2011). In conclusion,

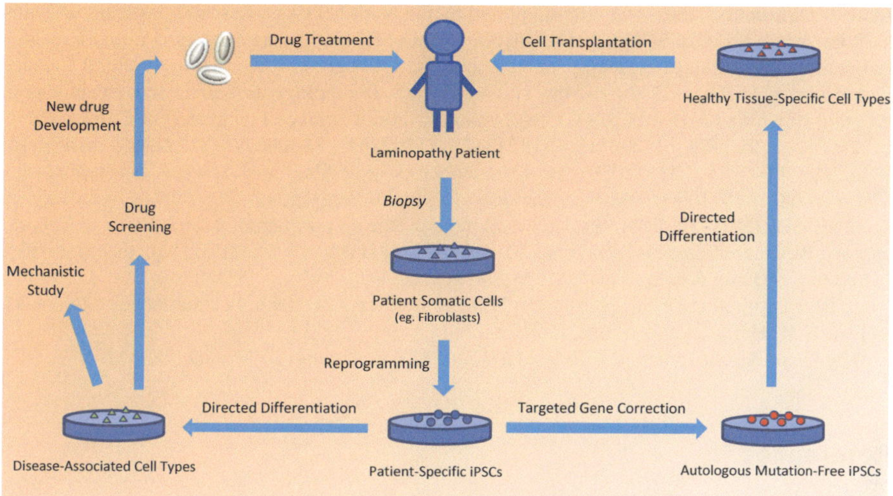

Fig. 5.2 Strategies of using induced pluripotent stem cell technology to study and treat laminopathies

the safety and efficiency of gene-editing technology will significantly speed up the development of disease modeling and clinical applications.

5.4.4 Perspectives

The main purpose of modeling laminopathies is to understand, diagnose, and eventually find treatments for these severe genetic disorders. Since the discovery of iPSC technologies, great advances have contributed to our understanding of the pathogenic mechanisms involved in many of these diseases. In addition, iPSC disease models possess a great potential for drug screening and autologous cell transplantation that may represent one of the future strategies of regenerative medicine. In our opinion, many laminopathies affecting heart, bone, or muscle are the result of gain- or loss-of-function mutations affecting tissue-specific cells, thus transplantation of safely corrected autologous cells might be a potential strategy to rescue disease phenotypes. However, screening drugs based on iPSC models is more feasible for premature aging diseases such as HGPS, because cellular treatment is unavailable considering HGPS is extremely hard to manage due to its early and severe symptoms. For instance, the commonly used farnesyltransferase inhibitors (FIT) could be tested on an HGPS model to study their therapeutic potential (Worman 2010). Hopefully, the fast-advancing field of disease modeling, directed differentiation, and targeted gene editing will eventually lead to better understanding and treatment of laminopathies (Fig. 5.2).

Acknowledgments This work was supported by the National Basic Research Program of China (973 Program, 2015CB964800; 2014CB910503; 2014CB964600), the Strategic Priority Research Program of the Chinese Academy of Sciences (XDA01020312), the Beijing Natural Science Foundation (7141005; 5142016), the National High Technology Research and Development Program of China (2015AA020307), the National Natural Science Foundation of China (NSFC: 81330008; 31222039; 31201111; 81371342; 81300261; 81300677; 81271266; 81471414; 81422017; 81401159; 31322039), the Program of Beijing Municipal Science and Technology Commission (Z151100003915072), the Key Research Program of the Chinese Academy of Sciences (KJZDEW-TZ-L05), the Thousand Young Talents program of China, National Laboratory of Biomacromolecules (012kf02;2013kf05;2013kf11;2014kf02;2015kf10), the State Key Laboratory of Drug Research (SIMM1302KF-17).

JCIB was supported by G. Harold and Leila Y. Mathers Charitable Foundation, The Leona M. and Harry B. Helmsley Charitable Trust.

The Glenn Foundation and Universidad Católica San Antonio de Murcia (UCAM).

References

Andres V, Gonzalez JM (2009) Role of A-type lamins in signaling, transcription, and chromatin organization. J Cell Biol 187:945–957

Burtner CR, Kennedy BK (2010) Progeria syndromes and ageing: what is the connection? Nat Rev Mol Cell Biol 11:567–578

Butin-Israeli V, Adam SA, Goldman AE, Goldman RD (2012) Nuclear lamin functions and disease. Trends Genet 28:464–471

Coffinier C, Chang SY, Nobumori C, Tu Y, Farber EA, Toth JI, Fong LG, Young SG (2010) Abnormal development of the cerebral cortex and cerebellum in the setting of lamin B2 deficiency. Proc Natl Acad Sci U S A 107:5076–5081

Coffinier C, Jung HJ, Nobumori C, Chang S, Tu Y, Barnes RH 2nd, Yoshinaga Y, de Jong PJ, Vergnes L, Reue K, Fong LG, Young SG (2011) Deficiencies in lamin B1 and lamin B2 cause neurodevelopmental defects and distinct nuclear shape abnormalities in neurons. Mol Biol Cell 22:4683–4693

Constantinescu D, Gray HL, Sammak PJ, Schatten GP, Csoka AB (2006) Lamin A/C expression is a marker of mouse and human embryonic stem cell differentiation. Stem Cells 24:177–185

Cristancho AG, Lazar MA (2011) Forming functional fat: a growing understanding of adipocyte differentiation. Nat Rev Mol Cell Biol 12:722–734

Dechat T, Pfleghaar K, Sengupta K, Shimi T, Shumaker DK, Solimando L, Goldman RD (2008) Nuclear lamins: major factors in the structural organization and function of the nucleus and chromatin. Genes Dev 22:832–853

Dechat T, Adam SA, Taimen P, Shimi T, Goldman RD (2010) Nuclear lamins. Cold Spring Harb Perspect Biol 2:a000547

Fatkin D, MacRae C, Sasaki T, Wolff MR, Porcu M, Frenneaux M, Atherton J, Vidaillet HJ Jr, Spudich S, De Girolami U, Seidman JG, Seidman C, Muntoni F, Muehle G, Johnson W, McDonough B (1999) Missense mutations in the rod domain of the lamin A/C gene as causes of dilated cardiomyopathy and conduction-system disease. N Engl J Med 341:1715–1724

Gerace L, Blum A, Blobel G (1978) Immunocytochemical localization of the major polypeptides of the nuclear pore complex-lamina fraction. Interphase and mitotic distribution. J Cell Biol 79:546–566

Gomez-Cavazos JS, Hetzer MW (2012) Outfits for different occasions: tissue-specific roles of nuclear envelope proteins. Curr Opin Cell Biol 24:775–783

Gorenne I, Kavurma M, Scott S, Bennett M (2006) Vascular smooth muscle cell senescence in atherosclerosis. Cardiovasc Res 72:9–17

Goulburn AL, Alden D, Davis RP, Micallef SJ, Ng ES, Yu QC, Lim SM, Soh CL, Elliott DA, Hatzistavrou T, Bourke J, Watmuff B, Lang RJ, Haynes JM, Pouton CW, Giudice A,

Trounson AO, Anderson SA, Stanley EG, Elefanty AG (2011) A targeted NKX2.1 human embryonic stem cell reporter line enables identification of human basal forebrain derivatives. Stem Cells 29:462–473

Handel EM, Cathomen T (2011) Zinc-finger nuclease based genome surgery: it's all about specificity. Curr Gene Ther 11:28–37

Hegele RA, Cao H, Liu DM, Costain GA, Charlton-Menys V, Rodger NW, Durrington PN (2006) Sequencing of the reannotated LMNB2 gene reveals novel mutations in patients with acquired partial lipodystrophy. Am J Hum Genet 79:383–389

Ho JC, Zhou T, Lai WH, Huang Y, Chan YC, Li X, Wong NL, Li Y, Au KW, Guo D, Xu J, Siu CW, Pei D, Tse HF, Esteban MA (2011) Generation of induced pluripotent stem cell lines from 3 distinct laminopathies bearing heterogeneous mutations in lamin A/C. Aging 3:380–390

Hockemeyer D, Wang H, Kiani S, Lai CS, Gao Q, Cassady JP, Cost GJ, Zhang L, Santiago Y, Miller JC, Zeitler B, Cherone JM, Meng X, Hinkley SJ, Rebar EJ, Gregory PD, Urnov FD, Jaenisch R (2011) Genetic engineering of human pluripotent cells using TALE nucleases. Nat Biotechnol 29:731–734

Jung HJ, Coffinier C, Choe Y, Beigneux AP, Davies BS, Yang SH, Barnes RH 2nd, Hong J, Sun T, Pleasure SJ, Young SG, Fong LG (2012) Regulation of prelamin A but not lamin C by miR-9, a brain-specific microRNA. Proc Natl Acad Sci U S A 109:E423–E431

Jung HJ, Nobumori C, Goulbourne CN, Tu Y, Lee JM, Tatar A, Wu D, Yoshinaga Y, de Jong PJ, Coffinier C, Fong LG, Young SG (2013) Farnesylation of lamin B1 is important for retention of nuclear chromatin during neuronal migration. Proc Natl Acad Sci U S A 110:E1923–E1932

Kudlow BA, Kennedy BK, Monnat RJ Jr (2007) Werner and Hutchinson-Gilford progeria syndromes: mechanistic basis of human progeroid diseases. Nat Rev Mol Cell Biol 8:394–404

Liu GH, Barkho BZ, Ruiz S, Diep D, Qu J, Yang SL, Panopoulos AD, Suzuki K, Kurian L, Walsh C, Thompson J, Boue S, Fung HL, Sancho-Martinez I, Zhang K, Yates J 3rd, Izpisua Belmonte JC (2011a) Recapitulation of premature ageing with iPSCs from Hutchinson-Gilford progeria syndrome. Nature 472:221–225

Liu GH, Suzuki K, Qu J, Sancho-Martinez I, Yi F, Li M, Kumar S, Nivet E, Kim J, Soligalla RD, Dubova I, Goebl A, Plongthongkum N, Fung HL, Zhang K, Loring JF, Laurent LC, Izpisua Belmonte JC (2011b) Targeted gene correction of laminopathy-associated LMNA mutations in patient-specific iPSCs. Cell Stem Cell 8:688–694

Liu GH, Qu J, Suzuki K, Nivet E, Li M, Montserrat N, Yi F, Xu X, Ruiz S, Zhang W, Wagner U, Kim A, Ren B, Li Y, Goebl A, Kim J, Soligalla RD, Dubova I, Thompson J, Yates J 3rd, Esteban CR, Sancho-Martinez I, Izpisua Belmonte JC (2012) Progressive degeneration of human neural stem cells caused by pathogenic LRRK2. Nature 491:603–607

Mali P, Yang L, Esvelt KM, Aach J, Guell M, DiCarlo JE, Norville JE, Church GM (2013) RNA-guided human genome engineering via Cas9. Science 339:823–826

Merideth MA, Gordon LB, Clauss S, Sachdev V, Smith AC, Perry MB, Brewer CC, Zalewski C, Kim HJ, Solomon B, Brooks BP, Gerber LH, Turner ML, Domingo DL, Hart TC, Graf J, Reynolds JC, Gropman A, Yanovski JA, Gerhard-Herman M, Collins FS, Nabel EG, Cannon RO 3rd, Gahl WA, Introne WJ (2008) Phenotype and course of Hutchinson-Gilford progeria syndrome. N Engl J Med 358:592–604

Mounkes LC, Stewart CL (2004) Aging and nuclear organization: lamins and progeria. Curr Opin Cell Biol 16:322–327

Nagano A, Koga R, Ogawa M, Kurano Y, Kawada J, Okada R, Hayashi YK, Tsukahara T, Arahata K (1996) Emerin deficiency at the nuclear membrane in patients with Emery-Dreifuss muscular dystrophy. Nat Genet 12:254–259

Nakamura K, Nakano S, Miyoshi T, Yamanouchi K, Matsuwaki T, Nishihara M (2012) Age-related resistance of skeletal muscle-derived progenitor cells to SPARC may explain a shift from myogenesis to adipogenesis. Aging 4:40–48

Niedernhofer LJ, Glorioso JC, Robbins PD (2011) Dedifferentiation rescues senescence of progeria cells but only while pluripotent. Stem Cell Res Ther 2:28

Nissan X, Blondel S, Navarro C, Maury Y, Denis C, Girard M, Martinat C, De Sandre-Giovannoli-A, Levy N, Peschanski M (2012) Unique preservation of neural cells in Hutchinson- Gilford progeria syndrome is due to the expression of the neural-specific miR-9 microRNA. Cell Rep 2:1–9

Olive M, Harten I, Mitchell R, Beers JK, Djabali K, Cao K, Erdos MR, Blair C, Funke B, Smoot L, Gerhard-Herman M, Machan JT, Kutys R, Virmani R, Collins FS, Wight TN, Nabel EG, Gordon LB (2010) Cardiovascular pathology in Hutchinson-Gilford progeria: correlation with the vascular pathology of aging. Arterioscler Thromb Vasc Biol 30:2301–2309

Padiath QS, Saigoh K, Schiffmann R, Asahara H, Yamada T, Koeppen A, Hogan K, Ptacek LJ, Fu YH (2006) Lamin B1 duplications cause autosomal dominant leukodystrophy. Nat Genet 38: 1114–1123

Pan H, Zhang W, Liu GH (2011) Find and replace: editing human genome in pluripotent stem cells. Protein Cell 2:950–956

Rana TM (2007) Illuminating the silence: understanding the structure and function of small RNAs. Nat Rev Mol Cell Biol 8:23–36

Rosen ED, MacDougald OA (2006) Adipocyte differentiation from the inside out. Nat Rev Mol Cell Biol 7:885–896

Ruby KM, Zheng B (2009) Gene targeting in a HUES line of human embryonic stem cells via electroporation. Stem Cells 27:1496–1506

Scaffidi P, Misteli T (2005) Reversal of the cellular phenotype in the premature aging disease Hutchinson-Gilford progeria syndrome. Nat Med 11:440–445

Scaffidi P, Misteli T (2006) Lamin A-dependent nuclear defects in human aging. Science 312: 1059–1063

Schreiber KH, Kennedy BK (2013) When lamins go bad: nuclear structure and disease. Cell 152:1365–1375

Siu CW, Lee YK, Ho JC, Lai WH, Chan YC, Ng KM, Wong LY, Au KW, Lau YM, Zhang J, Lay KW, Colman A, Tse HF (2012) Modeling of lamin A/C mutation premature cardiac aging using patient-specific induced pluripotent stem cells. Aging 4:803–822

Soldner F, Laganiere J, Cheng AW, Hockemeyer D, Gao Q, Alagappan R, Khurana V, Golbe LI, Myers RH, Lindquist S, Zhang L, Guschin D, Fong LK, Vu BJ, Meng X, Urnov FD, Rebar EJ, Gregory PD, Zhang HS, Jaenisch R (2011) Generation of isogenic pluripotent stem cells differing exclusively at two early onset Parkinson point mutations. Cell 146:318–331

Suzuki K, Mitsui K, Aizawa E, Hasegawa K, Kawase E, Yamagishi T, Shimizu Y, Suemori H, Nakatsuji N, Mitani K (2008) Highly efficient transient gene expression and gene targeting in primate embryonic stem cells with helper-dependent adenoviral vectors. Proc Natl Acad Sci U S A 105:13781–13786

Vergnes L, Peterfy M, Bergo MO, Young SG, Reue K (2004) Lamin B1 is required for mouse development and nuclear integrity. Proc Natl Acad Sci U S A 101:10428–10433

Worman HJ (2010) Prelamin A prenylation and the treatment of progeria. J Lipid Res 51:223–225

Xiong ZM, LaDana C, Wu D, Cao K (2013) An inhibitory role of progerin in the gene induction network of adipocyte differentiation from iPS cells. Aging 5:288–303

Xue H, Wu S, Papadeas ST, Spusta S, Swistowska AM, MacArthur CC, Mattson MP, Maragakis NJ, Capecchi MR, Rao MS, Zeng X, Liu Y (2009) A targeted neuroglial reporter line generated by homologous recombination in human embryonic stem cells. Stem Cells 27:1836–1846

Yamanaka S, Blau HM (2010) Nuclear reprogramming to a pluripotent state by three approaches. Nature 465:704–712

Young SG, Meta M, Yang SH, Fong LG (2006) Prelamin A farnesylation and progeroid syndromes. J Biol Chem 281:39741–39745

Young SG, Jung HJ, Coffinier C, Fong LG (2012) Understanding the roles of nuclear A- and B-type lamins in brain development. J Biol Chem 287:16103–16110

Zhang Q, Bethmann C, Worth NF, Davies JD, Wasner C, Feuer A, Ragnauth CD, Yi Q, Mellad JA, Warren DT, Wheeler MA, Ellis JA, Skepper JN, Vorgerd M, Schlotter-Weigel B, Weissberg PL, Roberts RG, Wehnert M, Shanahan CM (2007) Nesprin-1 and -2 are involved in the pathogenesis of Emery Dreifuss muscular dystrophy and are critical for nuclear envelope integrity. Hum Mol Genet 16:2816–2833

Zhang J, Lian Q, Zhu G, Zhou F, Sui L, Tan C, Mutalif RA, Navasankari R, Zhang Y, Tse HF, Stewart CL, Colman A (2011) A human iPSC model of Hutchinson Gilford Progeria reveals vascular smooth muscle and mesenchymal stem cell defects. Cell Stem Cell 8:31–45

Zou J, Maeder ML, Mali P, Pruett-Miller SM, Thibodeau-Beganny S, Chou BK, Chen G, Ye Z, Park IH, Daley GQ, Porteus MH, Joung JK, Cheng L (2009) Gene targeting of a disease-related gene in human induced pluripotent stem and embryonic stem cells. Cell Stem Cell 5:97–110

Hematological Disorders

6

Megumu K. Saito and Akira Niwa

Abstract

Hematological and immunological disorders are abnormalities of the blood system. Although the elucidation of their cellular pathophysiology has been largely based on in vitro studies using patient-derived primary hematopoietic cells or animal models, these approaches have potential limitations. Induced pluripotent stem cells (iPSCs) are potential cell sources for regenerative medicine and other clinical applications, such as cell therapy, drug screening, and the investigation of disease mechanisms. The discovery of disease-associated iPSCs has led to the development of a new field of disease modeling, which can provide somatic cells that cannot be directly obtained from the patient. There have been a number of reports evaluating the use of patient-derived iPSCs to treat hematological and immunological disorders, such as bone marrow failure, hemoglobinopathy, congenital immunodeficiency, hematological malignancies, and chromosomal abnormalities. In this chapter, we review these reports and focus on the application of disease-associated iPSCs for understanding human hematological and immunological disorders while discussing the current state of hematopoietic differentiation and future perspectives.

Keywords

Hematopoietic stem cell differentiation • Bone marrow failure • Hemoglobinopathy • Thalassemia • Leukemia • Immunodeficiency • Autoinflammatory syndrome

M.K. Saito (✉) • A. Niwa
Department of Clinical Application, Center for iPS cell research and application, Kyoto University, 53 Shogoin-Kawahara, Sakyo-ku, Kyoto 606-8507, Japan
e-mail: msaito@cira.kyoto-u.ac.jp

K. Fukuda (ed.), *Human iPS Cells in Disease Modelling*,
DOI 10.1007/978-4-431-55966-5_6

6.1 Introduction

Hematological and immunological disorders are abnormalities of the blood system. Although the elucidation of their cellular pathophysiology has been largely based on in vitro studies using patient-derived primary hematopoietic cells or animal models, these approaches have potential limitations. For example, patient-derived cells cannot be obtained in unlimited quantities, and their in vitro functions can be affected by in vivo conditions, such as the cytokine milieu or presence of therapeutic agents. Furthermore, in cases involving multisystem disorders, such as those associated with chromosomal abnormalities (Bruwier and Chantrain 2012), defects in DNA repair (Deans and West 2011) or metabolic disorders (Stoffels and Simon 2011), obtaining patient-derived samples other than blood is difficult, hampering the analysis of affected cells and tissues. On the other hand, although murine models have provided important insights into various disorders, differences in hematological and immunological development between mice and humans sometimes cause discrepancies in the resulting phenotypes.

Due to their pluripotency and capacity for self-renewal, human pluripotent stem cells (PSCs), such as embryonic stem cells (ESCs) and induced pluripotent stem cells (iPSCs), are potential sources of cells for use in regenerative medicine and other clinical applications, such as cell therapy, drug screening, toxicology testing, and the investigation of disease mechanisms (Thomson et al. 1998; Keller 2005). iPSCs, which were first established by Takahashi and Yamanaka, are reprogrammed somatic cells with ESC-like characteristics that are generated by introducing certain transcriptional factors, such as *OCT3/4*, *SOX2*, *KLF4*, and *c-MYC* (Takahashi et al. 2007; Takahashi and Yamanaka 2006). These transcriptional factors can be delivered into the source cells using viral vectors, episomal vectors (Okita et al. 2011), piggyBac transposons (Woltjen et al. 2009), or modified synthetic RNA (Warren et al. 2010). The discovery of disease-associated iPSCs has led to the development of a new field of disease modeling, which can provide somatic cells that cannot be directly obtained from the patient.

6.2 Directed Differentiation into Hematopoietic Stem Cells from Human PSCs

Although patient- and/or disease-specific iPSCs are an important resource for unraveling human hematological disorders, the development of a robust and simple hematopoietic differentiation system that can reliably mimic in vivo hematopoiesis is necessary for this purpose. The leading methods of hematopoietic cell induction from PSCs employ two different systems: namely, monolayer animal-derived stromal cell co-culture and three-dimensional embryoid body (EB) formation. Both methods can be used to produce hematopoietic cells from mesodermal progenitors, and combinations of cytokines can control, at least to some extent, the specific lineage commitment (Umeda et al. 2004, 2006; Ji et al. 2008; Vodyanik et al. 2005; Kitajima et al. 2006; Takayama et al. 2008; Choi et al. 2009; Niwa et al. 2009; Timmermans et al. 2009). We recently established a robust and simple monolayer hematopoietic cell differentiation system using human PSCs (Niwa et al. 2011; Yanagimachi et al. 2013). Our system is free from xeno-feeder cells and serum and can be used to trace the in vitro differentiation of human PSCs into multiple lineages of definitive blood cells, such as functional erythrocytes and neutrophils. We also developed a robust and highly efficient method of differentiating functional monocytic cells from ESCs/iPSCs under serum- and feeder cell-free conditions (Yanagimachi et al. 2013). This method repeatedly produces floating monocytes during each course of differentiation. Pluripotent stem cell-derived monocytes can be differentiated into both functional macrophages and dendritic cells.

Because human PSCs are feasible cell sources for use in various clinical applications, the scientific and medical communities have shown continuing interest in hematopoietic stem cell (HSC) induction from PSCs. Previous trials have indicated that murine ESC-derived hematopoietic cells overexpressing *HoxB4* induce long-term myelo-lymphoid reconstitution in the bone marrow of lethally irradiated recipient mice (Kyba et al. 2002). However, the transduction of *HOXB4* is not sufficient to induce the development of fully functional human HSCs (Schiedlmeier et al. 2003), and it remains a challenge to develop *bona fide* human HSCs with bone marrow reconstitution activity at the single-cell level.

Despite recent advances, the directed differentiation of human PSCs into definitive hematopoietic cells in vitro also remains challenging. Most cultures primarily develop into nucleated erythrocytes with a primitive or definitive fatal type of hemoglobin expression pattern (α- and γ-globins), and the robust and effective derivation of enucleated adult type α- and β-globin-expressing red blood cells from human PSCs remains elusive. Differentiating lymphoid cells is also relatively difficult. While T lymphocytes can be derived from human PSCs on OP9-DL1 feeder layers (Timmermans et al. 2009), the terminal differentiation into B lymphocytes remains to be accomplished, as it has thus far not been possible to induce cells to go through the pre-B state (Carpenter et al. 2011). Several groups have reported the derivation of functional natural killer cells from PSCs (Vodyanik et al. 2005; Ni et al. 2011).

6.3 Disease-Associated iPS Cells Obtained from Patients with Hematological and Immunological Disorders

A number of disease-associated iPSCs generated from patients with hematological and/or immunological disorders have been reported (Table 6.1). Furthermore, several papers recently reported the establishment of iPSC clones while discussing the potential usefulness of disease-associated iPSCs as a resource for disease analyses. In this section, we focus on reports that describe the in vitro phenotypes of hematopoietic cells derived from disease-specific iPSCs.

Table 6.1 Reported disease-associated iPS cell lines representing hematological and immunological disorders

Category	Name of disorder	Molecular defect	Author	Year
Bone marrow failure	Fanconi anemia	*FANCA/FANCC*	Muller	2012
Bone marrow failure	Fanconi anemia	*FANCA/FANCD2*	Raya	2009
Bone marrow failure	Shwachman-Bodian-Diamond syndrome	*SBDS*	Park	2008
Bone marrow failure	Dyskeratosis congenita	*DKC*	Batista	2011
Chromosomal abnormality	Down syndrome	Trisomy 21	Park	2008
Hemoglobinopathy	Sickle cell anemia	*HBB*	Sebastiano	2011
Hemoglobinopathy	Sickle cell anemia	*HBB*	Mali[a]	2008
Hemoglobinopathy	Sickle cell anemia	*HBB*	Zou[a]	2012
Hereditary anemia	Beta-thalassemia	*HBB*	Ye	2009
Hereditary anemia	Beta-thalassemia	*HBB*	Wang[b]	2009
Hereditary anemia	Beta-thalassemia	*HBB*	Papapetrou	2011
Hereditary anemia	Beta-thalassemia	*HBB*	Wang[b]	2012
Hereditary anemia	Beta-thalassemia	*HBB*	Fan	2012
Hematological malignancy	Chronic myeloid leukemia	*BCR/ABL*	Hu	2011
Hematological malignancy	Chronic myeloid leukemia	*BCR/ABL*	Carette	2011
Hematological malignancy	Chronic myeloid leukemia	*BCR/ABL*	Kumano	2012
Rare blood type	Rare ABO blood type (Bombay)	*FUT1/FUT2*	Seifinejad	2009
Immunodeficiency	Chronic granulomatous disease	*p47phox, qp91phox*	Jiang	2012
Immunodeficiency	Chronic granulomatous disease	*gp91phox*	Zou	2011
Immunodeficiency	Severe combined immunodeficiency	*ADA*	Park	2008
Immunodeficiency	Severe combined immunodeficiency	*RAG1*	Pessach	2010
Immunodeficiency	Omenn syndrome	*RAG1*	Pessach	2010
Immunodeficiency	Herpes simplex type 1 encephalitis	*STAT1*	Pessach	2010
Immunodeficiency	Herpes simplex type 2 encephalitis	*LR3*	Pessach	2010
Immunodeficiency	Cartilage hair hypoplasia	*RMRP*	Pessach	2010
Autoinflammatory syndrome	CINCA syndrome	*NLRP3*	Tanaka	2012

[a,b]Reported by the same group

6.3.1 Bone Marrow Failure

The first report of disease-associated iPSCs derived from patients with hematological disorders involved iPSCs obtained from a patient with Fanconi anemia (FA) (Raya et al. 2009), an autosomal recessive disorder of pediatric bone marrow failure. FA is a genetically heterogeneous syndrome caused by a defect in a series of FANC family genes responsible for DNA repair. Raya et al. established FA patient-derived iPSC clones from FANCA- and FANCD2-deficient patients after correcting genetic defects in the parental fibroblasts. The uncorrected fibroblasts could not be reprogrammed into iPSCs, indicating that the Fanconi anemia pathway is requisite for reprogramming. The corrected iPSC clones were able to differentiate into hematopoietic cells normally. Recently, another group succeeded in generating iPSCs obtained from FANCA- and FANCC-deficient patients without complementation under hypoxic conditions; however, the efficiency was extremely low (Muller et al. 2012). Regarding hematological phenotypes, complimented iPSCs exhibit a differentiation efficacy equivalent to that of normal donor-derived iPSCs. Another group reported that they succeeded in reprogramming FANCA-, FANCC-, and FANCD2-deficient fibroblasts, although only the FANCC-iPSCs could be maintained (Yung et al. 2013). These works provide unique insights into the functions of the FANC pathway in reprogramming and early hematopoietic development.

Shwachman-Diamond syndrome (SDS) is a rare autosomal recessive disorder characterized by exocrine pancreatic insufficiency and hematopoietic dysfunction (Myers et al. 2013). It is caused by the loss of the Shwachman-Bodian-Diamond syndrome (SBDS) gene. Tulpule et al. generated both SDS patient-derived iPSCs and SBDS-knockdown ESCs and showed that protease-mediated autodigestion is linked to pancreatic and hematopoietic phenotypes, such as stress-mediated cell death (Tulpule et al. 2013). Despite the phenotypic variability in SBDS-deficient lines, restoring the SBDS protein expression reverses the phenotypes. These results at least partly prove the existence of a genotype-phenotype relationship in this disease model using ESC/iPSCs.

6.3.2 Hemoglobinopathies

For β-thalassemia, one of the most common hereditary forms of anemia (Weatherall and Clegg 2001), disease-associated iPSC studies have been primarily conducted toward proving that the iPSC technology can be used to generate gene-corrected cells with potential value for cell therapy. Wang et al. genetically corrected iPSCs obtained from a β-thalassemia patient using homologous recombination and differentiated the cells into hematopoietic progenitors (Wang et al. 2009, 2012). When the cells were transferred to sublethally irradiated NOD/SCID mice, the hemoglobin levels recovered efficiently. Papapetrou et al. showed that the transduction of a lentivirally encoded β-globin transgene into genomic safe harbors enables the high expression of the transgene in β-thalassemia iPSC-derived erythroid progenitors (Papapetrou et al. 2011). Similar to that observed for β-thalassemia, two groups reported genetic correction using cells obtained from patients with sickle cell anemia (Zou et al. 2011a; Sebastiano et al. 2011; Mali et al. 2008).

6.3.3 Primary Immunodeficiency

Two groups established iPSCs obtained from patients with chronic granulomatous disease (CGD) (Zou et al. 2011b; Jiang et al. 2012), a primary immunodeficiency characterized by the impaired phagocytic killing of microorganisms by neutrophils and macrophages (Winkelstein et al. 2000). Both groups demonstrated that the differentiated neutrophils derived from the disease-associated iPSCs do not produce reactive oxygen species (ROS) in response to proper stimuli. As a model for gene therapy, Zou et al. restored the neutrophil ROS production in X-linked CGD iPSCs using zinc finger nuclease-mediated gene targeting of a single-copy gp91phox therapeutic minigene into one allele of the "safe harbor" *AAVS1* locus (Zou et al. 2011b).

Severe congenital neutropenia (SCN) refers to a group of heterogeneous syndromes characterized by severe neutropenia at birth. Past studies have accumulated a list of associated genes, including HAX1, ELANE, GFI1, WAS, and G6PC3 (Bouma et al. 2010). Two Japanese groups independently described iPSCs obtained from patients with congenital neutropenia syndromes (Hiramoto et al. 2013; Morishima et al. 2013). Hiramoto et al. established iPSCs obtained from an individual with SCN who had a heterozygous mutation in the neutrophil elastase *ELANE* gene. ELANE-mutated iPSCs recapitulate neutrophil maturation arrest refractory to stimulation with granulocyte-colony stimulating factor (G-CSF). $CD34^+$ hematopoietic progenitors derived from ELANE-mutated iPSCs exhibit comparable amounts of erythroid and mixed-lineage hematopoietic colonies, while the number of granulocyte and granulocyte-macrophage colonies is significantly decreased. The authors found the Wnt3a/β-catenin pathway is downregulated in ELANE-mutated iPSC-derived neutrophils and that the administration of Wnt3a rescues the maturation of ELANE-mutated iPSC-derived neutrophils, suggesting a potential therapeutic approach for treating SCN with ELANE mutations. Morishima et al. established iPSCs obtained from a SCN patient with a homozygous mutation of the *HAX1* gene (Morishima et al. 2013). Like ELANE-mutated SCN-iPSCs, HAX1-mutated iPSC-derived hematopoietic progenitors exhibit reduced clonogenicity of granulocyte and granulocyte-macrophage colonies, with neutrophil maturation arrest recapitulated in vitro.

6.3.4 Malignant Hematopoietic Disorders

Disease-associated iPSCs obtained from patients with chronic myeloid leukemia (CML) have also been reported (Hu et al. 2011; Carette et al. 2010; Kumano et al. 2012). The sources of iPSCs include a cell line, KBM7

(Carette et al. 2010), primary bone marrow cells (Hu et al. 2011), and $CD34^+$ cells (Kumano et al. 2012). All iPSC clones exhibit the translocation of 9;22 breakpoints of the *BCR/ABL* fusion gene. Interestingly, although the parental cell lines are sensitive to the tyrosine kinase inhibitor imatinib, thus showing their dependency on *BCR/ABL* oncogene signaling, a loss of oncogene addiction was reported in the reprogrammed iPSC clones (Kumano et al. 2012). The hematopoietic differentiation of these iPSCs recovered their sensitivity to imatinib.

Juvenile myelomonocytic leukemia (JMML) is a severe pediatric myeloproliferative disorder with a poor prognosis (Niemeyer and Kratz 2008). iPSCs have been established from the circulating malignant cells of two JMML patients with a somatic heterozygous mutation of E76K in the *PTPN11* gene (Gandre-Babbe et al. 2013). The hematopoietic progenitor cells derived from these JMML-iPSCs exhibit increased myeloid clonogenicity with an increased frequency of myeloid progenitors during directed differentiation. In contrast to that observed in SCN-iPSCs, JMML-iPSC-derived hematopoietic progenitors demonstrate constitutive hyperactivation of the G-CSF signaling pathway, similar to primary JMML blasts. The authors evaluated the utility of JMML-iPSCs for use in drug screening by testing the effects of inhibitors of MEK kinase and found that MEK inhibition normalizes the GM-CSF dose response and degree of myeloid clonogenicity.

6.3.5 Hematopoietic Abnormalities Associated with Down Syndrome

Down syndrome (DS), also known as trisomy 21, is the most common autosomal chromosomal disorder, with various clinical symptoms, including heart defects, mental retardation, and premature aging. Approximately 10 % of neonates with DS develop characteristic clonal preleukemia, termed transient abnormal myelopoiesis or transient myeloproliferative disease, which progresses to acute megakaryoblastic leukemia in 10–30 % of TAM patients (Webb et al. 2007; Kurahashi et al. 1992). Because the initial onset of TAM/TMD during fatal hematopoiesis and hematopoietic differentiation from iPSCs is usually limited to the primitive or definitive fatal stage, iPSC technology is suitable for studying the pathogenesis of TAM/TMD. Since DS patients develop both mental retardation and early-onset Alzheimer's disease during adulthood, studies of DS-iPSCs have primarily focused on functional or developmental deficits in pluripotent stem cell-derived neurons (Weick et al. 2013; Briggs et al. 2013; Shi et al. 2012; Lu et al. 2013). As for hematological disorders, two groups reported the in vitro hematological phenotypes of DS-iPSCs (Chou et al. 2012; Maclean et al. 2012). MacLean et al. obtained disomic and trisomic subclones derived from DS-iPSCs and used disomic clones as an isogenic control. They found that trisomic ES and/or iPS cells exhibit an increased frequency of both $CD43^+/CD235^+$ erythroid progenitors and $CD43^+CD235^-$ hematopoietic progenitors. The clonogenicity of unsorted embryonic body cells differentiated from trisomy clones is also increased in both myeloid and erythroid colonies. Chou et al. also developed iPSCs obtained from four DS individuals differentiated into hematopoietic lineages. They found that DS-iPSCs show a propensity toward differentiation into $CD41^{-}CD235^{+}$ erythroid progenitors, while the frequency of myeloid progenitors is decreased. The authors speculated that this distinct impact of trisomy 21 on myeloid differentiation may result from differences in the developmental stage of iPSC-derived hematopoietic disorders.

6.3.6 Dissecting Somatic Mutations with iPS Cells

One of the interesting characteristics of iPSCs is that each iPSC clone originates from a single somatic cell (Hanna et al. 2008). Taking advantage of this fact, we previously proposed that iPSC technology can be used to dissect and evaluate genetically different somatic cells obtained from individuals (Tanaka et al. 2012). In patients with CINCA syndrome, an autoinflammatory syndrome caused by mutations of the *NLRP3* gene, 30–40 % of cases involve mutations in *NLRP3* in only a small number of somatic cells (Saito et al. 2005, 2008), and it remains controversial whether the small fraction of *NLRP3*-mutated cells actually causes the strong autoinflammatory reactions observed in these patients or if all cells carry an unknown mutation of another gene that causes the disease. To resolve this controversy, mutant and nonmutant iPSC lines were established from CINCA patients with somatic mosaicism (Tanaka et al. 2012). By analyzing the disease-relevant characteristics of IL-1β secretion from iPSC-derived macrophages, we demonstrated that mutant macrophages are primarily responsible for the disease phenotype observed in mosaic patients, confirming the role of *NLRP3*.

6.4 Concluding Remarks

As discussed above, most previously reported disease-associated iPSC studies have been "disease-modeling" or proof-of-principle studies. In order to gain more insight into disease pathophysiology using iPSC technology, several issues still must be overcome. One of the most critical issues is to develop a method to obtain mature, fully functional hematopoietic cells, including HSCs. Another concern is that, even if iPSCs are obtained from individuals, the efficiency of differentiation and/or the functions of the hematopoietic cells will exhibit interclonal variation, which hampers the accurate estimation of the disease-associated phenotypes of patient-derived iPSCs. These variations may derive from interclonal genetic variation (Gore et al. 2011), epigenetic modification (Kim et al. 2010), the source of iPSCs, the presence of residual transgenes of each iPSC clone, or, in female cases, the alteration of the status of X chromosome inactivation (Mekhoubad et al. 2012; Marchetto et al. 2010). Additionally, fibroblasts obtained from patients with certain diseases, such as Fanconi anemia (Raya et al. 2009) and dyskeratosis congenita (Batista et al. 2011), demonstrate extremely low reprogramming efficiency. For these types of diseases, the development of a specific reprograming strategy, such as transient genetic complementation, is therefore required.

Although disease-associated iPSCs are useful tools, their proper differentiation into functional hematopoietic cells is essential for elucidating the cellular pathophysiology of hematopoietic and immunological diseases. The establishment of suitable disease models that can be used to represent in vivo phenotypes is also important. Rapid technological advances in iPSCs and their differentiation will open up a new horizon for studies that can aid in understanding human diseases.

Acknowledgments This work was partly supported from a grant from the Ministry of Education, Culture, Sports, Science and Technology (MEXT).

Conflicts of Interest The authors declare no conflicts of interest.

References

Batista LF, Pech MF, Zhong FL, Nguyen HN, Xie KT, Zaug AJ, Crary SM, Choi J, Sebastiano V, Cherry A, Giri N, Wernig M, Alter BP, Cech TR, Savage SA, Reijo Pera RA, Artandi SE (2011) Telomere shortening and loss of self-renewal in dyskeratosis congenita induced pluripotent stem cells. Nature 474(7351):399–402

Bouma G, Ancliff PJ, Thrasher AJ, Burns SO (2010) Recent advances in the understanding of genetic defects of neutrophil number and function. Br J Haematol 151(4):312–326. doi:10.1111/j.1365-2141.2010.08361.x

Briggs JA, Sun J, Shepherd J, Ovchinnikov DA, Chung TL, Nayler SP, Kao LP, Morrow CA, Thakar NY, Soo SY, Peura T, Grimmond S, Wolvetang EJ (2013) Integration-free induced pluripotent stem cells model genetic and neural developmental features of Down syndrome etiology. Stem Cells 31(3):467–478. doi:10.1002/stem.1297

Bruwier A, Chantrain CF (2012) Hematological disorders and leukemia in children with Down syndrome. Eur J Pediatr 171:1301–1307

Carette JE, Pruszak J, Varadarajan M, Blomen VA, Gokhale S, Camargo FD, Wernig M, Jaenisch R, Brummelkamp TR (2010) Generation of iPSCs from cultured human malignant cells. Blood 115(20):4039–4042

Carpenter L, Malladi R, Yang CT, French A, Pilkington KJ, Forsey RW, Sloane-Stanley J, Silk KM, Davies TJ, Fairchild PJ, Enver T, Watt SM (2011) Human induced pluripotent stem cells are capable of B-cell lymphopoiesis. Blood 117(15):4008–4011

Choi KD, Vodyanik MA, Slukvin II (2009) Generation of mature human myelomonocytic cells through expansion and differentiation of pluripotent stem cell-derived lin-CD34+CD43+CD45+ progenitors. J Clin Invest 119(9):2818–2829

Chou ST, Byrska-Bishop M, Tober JM, Yao Y, Vandorn D, Opalinska JB, Mills JA, Choi JK, Speck NA, Gadue P, Hardison RC, Nemiroff RL, French DL, Weiss MJ (2012) Trisomy 21-associated defects in human primitive hematopoiesis revealed through induced pluripotent stem cells. Proc Natl Acad Sci U S A 109(43):17573–17578. doi:10.1073/pnas.1211175109

Deans AJ, West SC (2011) DNA interstrand crosslink repair and cancer. Nat Rev Cancer 11(7):467–480

Gandre-Babbe S, Paluru P, Aribeana C, Chou ST, Bresolin S, Lu L, Sullivan SK, Tasian SK, Weng J, Favre H, Choi JK, French DL, Loh ML, Weiss MJ (2013) Patient-derived induced pluripotent stem cells recapitulate hematopoietic abnormalities of juvenile myelomonocytic leukemia. Blood 121(24):4925–4929. doi:10.1182/blood-2013-01-478412

Gore A, Li Z, Fung HL, Young JE, Agarwal S, Antosiewicz-Bourget J, Canto I, Giorgetti A, Israel MA, Kiskinis E, Lee JH, Loh YH, Manos PD, Montserrat N, Panopoulos AD, Ruiz S, Wilbert ML, Yu J, Kirkness EF, Izpisua Belmonte JC, Rossi DJ, Thomson JA, Eggan K,

Daley GQ, Goldstein LS, Zhang K (2011) Somatic coding mutations in human induced pluripotent stem cells. Nature 471(7336):63–67

Hanna J, Markoulaki S, Schorderet P, Carey BW, Beard C, Wernig M, Creyghton MP, Steine EJ, Cassady JP, Foreman R, Lengner CJ, Dausman JA, Jaenisch R (2008) Direct reprogramming of terminally differentiated mature B lymphocytes to pluripotency. Cell 133(2):250–264

Hiramoto T, Ebihara Y, Mizoguchi Y, Nakamura K, Yamaguchi K, Ueno K, Nariai N, Mochizuki S, Yamamoto S, Nagasaki M, Furukawa Y, Tani K, Nakauchi H, Kobayashi M, Tsuji K (2013) Wnt3a stimulates maturation of impaired neutrophils developed from severe congenital neutropenia patient-derived pluripotent stem cells. Proc Natl Acad Sci U S A 110(8):3023–3028. doi:10.1073/pnas.1217039110

Hu K, Yu J, Suknuntha K, Tian S, Montgomery K, Choi KD, Stewart R, Thomson JA, Slukvin II (2011) Efficient generation of transgene-free induced pluripotent stem cells from normal and neoplastic bone marrow and cord blood mononuclear cells. Blood 117(14):e109–e119

Ji P, Jayapal SR, Lodish HF (2008) Enucleation of cultured mouse fetal erythroblasts requires Rac GTPases and mDia2. Nat Cell Biol 10(3):314–321

Jiang Y, Cowley SA, Siler U, Melguizo D, Tilgner K, Browne C, Dewilton A, Przyborski S, Saretzki G, James WS, Seger RA, Reichenbach J, Lako M, Armstrong L (2012) Derivation and functional analysis of patient-specific induced pluripotent stem cells as an in vitro model of chronic granulomatous disease. Stem Cells 30(4):599–611

Keller G (2005) Embryonic stem cell differentiation: emergence of a new era in biology and medicine. Genes Dev 19(10):1129–1155

Kim K, Doi A, Wen B, Ng K, Zhao R, Cahan P, Kim J, Aryee MJ, Ji H, Ehrlich LI, Yabuuchi A, Takeuchi A, Cunniff KC, Hongguang H, McKinney-Freeman S, Naveiras O, Yoon TJ, Irizarry RA, Jung N, Seita J, Hanna J, Murakami P, Jaenisch R, Weissleder R, Orkin SH, Weissman IL, Feinberg AP, Daley GQ (2010) Epigenetic memory in induced pluripotent stem cells. Nature 467 (7313):285–290

Kitajima K, Tanaka M, Zheng J, Yen H, Sato A, Sugiyama D, Umehara H, Sakai E, Nakano T (2006) Redirecting differentiation of hematopoietic progenitors by a transcription factor, GATA-2. Blood 107(5):1857–1863

Kumano K, Arai S, Hosoi M, Taoka K, Takayama N, Otsu M, Nagae G, Ueda K, Nakazaki K, Kamikubo Y, Eto K, Aburatani H, Nakauchi H, Kurokawa M (2012) Generation of induced pluripotent stem cells from primary chronic myelogenous leukemia patient samples. Blood 119(26):6234–6242

Kurahashi H, Hara J, Yumura-Yagi K, Tawa A, Kawa-Ha K (1992) Transient abnormal myelopoiesis in Down's syndrome. Leuk Lymphoma 8(6):465–475. doi:10.3109/10428199209051029

Kyba M, Perlingeiro RC, Daley GQ (2002) HoxB4 confers definitive lymphoid-myeloid engraftment potential on embryonic stem cell and yolk sac hematopoietic progenitors. Cell 109(1): 29–37

Lu HE, Yang YC, Chen SM, Su HL, Huang PC, Tsai MS, Wang TH, Tseng CP, Hwang SM (2013) Modeling neurogenesis impairment in Down syndrome with induced pluripotent stem cells from Trisomy 21 amniotic fluid cells. Exp Cell Res 319(4):498–505. doi:10.1016/j.yexcr.2012.09.017

Maclean GA, Menne TF, Guo G, Sanchez DJ, Park IH, Daley GQ, Orkin SH (2012) Altered hematopoiesis in trisomy 21 as revealed through in vitro differentiation of isogenic human pluripotent cells. Proc Natl Acad Sci U S A 109(43):17567–17572. doi:10.1073/pnas.1215468109

Mali P, Ye Z, Hommond HH, Yu X, Lin J, Chen G, Zou J, Cheng L (2008) Improved efficiency and pace of generating induced pluripotent stem cells from human adult and fetal fibroblasts. Stem Cells 26(8):1998–2005

Marchetto MC, Carromeu C, Acab A, Yu D, Yeo GW, Mu Y, Chen G, Gage FH, Muotri AR (2010) A model for neural development and treatment of Rett syndrome using human induced pluripotent stem cells. Cell 143(4):527–539

Mekhoubad S, Bock C, de Boer AS, Kiskinis E, Meissner A, Eggan K (2012) Erosion of dosage compensation impacts human iPSC disease modeling. Cell Stem Cell 10(5):595–609

Morishima T, Watanabe KI, Niwa A, Hirai H, Saida S, Tanaka T, Kato I, Umeda K, Hiramatsu H, Saito MK, Matsubara K, Adachi S, Kobayashi M, Nakahata T, Heike T (2013) Genetic correction of HAX1 in induced pluripotent stem cells from a patient with severe congenital neutropenia improves defective granulopoiesis. Haematologica. doi:10.3324/haematol.2013.083873

Muller LU, Milsom MD, Harris CE, Vyas R, Brumme KM, Parmar K, Moreau LA, Schambach A, Park IH, London WB, Strait K, Schlaeger T, Devine AL, Grassman E, D'Andrea A, Daley GQ, Williams DA (2012) Overcoming reprogramming resistance of Fanconi anemia cells. Blood 119(23):5449–5457. doi:10.1182/blood-2012-02-408674

Myers KC, Davies SM, Shimamura A (2013) Clinical and molecular pathophysiology of Shwachman-Diamond syndrome: an update. Hematol Oncol Clin N Am 27(1):117–128, ix. doi:10.1016/j.hoc.2012.10.003

Ni Z, Knorr DA, Clouser CL, Hexum MK, Southern P, Mansky LM, Park IH, Kaufman DS (2011) Human pluripotent stem cells produce natural killer cells that mediate anti-HIV-1 activity by utilizing diverse cellular mechanisms. J Virol 85(1):43–50

Niemeyer CM, Kratz CP (2008) Paediatric myelodysplastic syndromes and juvenile myelomonocytic leukaemia: molecular classification and treatment options. Br J Haematol 140(6):610–624. doi:10.1111/j.1365-2141.2007.06958.x

Niwa A, Umeda K, Chang H, Saito M, Okita K, Takahashi K, Nakagawa M, Yamanaka S, Nakahata T, Heike T (2009) Orderly hematopoietic development of induced pluripotent stem cells via Flk-1(+) hemoangiogenic progenitors. J Cell Physiol 221(2):367–377. doi:10.1002/jcp.21864

Niwa A, Heike T, Umeda K, Oshima K, Kato I, Sakai H, Suemori H, Nakahata T, Saito MK (2011) A novel serum-free monolayer culture for orderly hematopoietic differentiation of human pluripotent cells via mesodermal progenitors. PLoS One 6(7), e22261. doi:10.1371/journal.pone.0022261

Okita K, Matsumura Y, Sato Y, Okada A, Morizane A, Okamoto S, Hong H, Nakagawa M, Tanabe K, Tezuka K, Shibata T, Kunisada T, Takahashi M, Takahashi J, Saji H, Yamanaka S (2011) A more efficient method to generate integration-free human iPS cells. Nat Methods 8(5):409–412. doi:10.1038/nmeth.1591

Papapetrou EP, Lee G, Malani N, Setty M, Riviere I, Tirunagari LM, Kadota K, Roth SL, Giardina P, Viale A, Leslie C, Bushman FD, Studer L, Sadelain M (2011) Genomic safe harbors permit high beta-globin transgene expression in thalassemia induced pluripotent stem cells. Nat Biotechnol 29(1):73–78

Raya A, Rodriguez-Piza I, Guenechea G, Vassena R, Navarro S, Barrero MJ, Consiglio A, Castella M, Rio P, Sleep E, Gonzalez F, Tiscornia G, Garreta E, Aasen T, Veiga A, Verma IM, Surralles J, Bueren J, Izpisua Belmonte JC (2009) Disease-corrected haematopoietic progenitors from Fanconi anaemia induced pluripotent stem cells. Nature 460(7251):53–59

Saito M, Fujisawa A, Nishikomori R, Kambe N, Nakata-Hizume M, Yoshimoto M, Ohmori K, Okafuji I, Yoshioka T, Kusunoki T, Miyachi Y, Heike T, Nakahata T (2005) Somatic mosaicism of CIAS1 in a patient with chronic infantile neurologic, cutaneous, articular syndrome. Arthritis Rheum 52(11):3579–3585. doi:10.1002/art.21404

Saito M, Nishikomori R, Kambe N, Fujisawa A, Tanizaki H, Takeichi K, Imagawa T, Iehara T, Takada H, Matsubayashi T, Tanaka H, Kawashima H, Kawakami K, Kagami S, Okafuji I, Yoshioka T, Adachi S, Heike T, Miyachi Y, Nakahata T (2008) Disease-associated CIAS1 mutations induce monocyte death, revealing low-level mosaicism in mutation-negative cryopyrin-associated periodic syndrome patients. Blood 111(4):2132–2141. doi:10.1182/blood-2007-06-094201

Schiedlmeier B, Klump H, Will E, Arman-Kalcek G, Li Z, Wang Z, Rimek A, Friel J, Baum C, Ostertag W (2003) High-level ectopic HOXB4 expression confers a profound in vivo competitive growth advantage on human cord blood CD34+ cells, but impairs lymphomyeloid differentiation. Blood 101(5):1759–1768

Sebastiano V, Maeder ML, Angstman JF, Haddad B, Khayter C, Yeo DT, Goodwin MJ, Hawkins JS, Ramirez CL, Batista LF, Artandi SE, Wernig M, Joung JK (2011) In situ genetic correction of the sickle cell anemia mutation in human induced pluripotent stem cells using engineered zinc finger nucleases. Stem Cells 29(11):1717–1726

Shi Y, Kirwan P, Smith J, MacLean G, Orkin SH, Livesey FJ (2012) A human stem cell model of early Alzheimer's disease pathology in Down syndrome. Sci Transl Med 4(124):124–129. doi:10.1126/scitranslmed.3003771

Stoffels M, Simon A (2011) Hyper-IgD syndrome or mevalonate kinase deficiency. Curr Opin Rheumatol 23(5):419–423

Takahashi K, Yamanaka S (2006) Induction of pluripotent stem cells from mouse embryonic and adult fibroblast cultures by defined factors. Cell 126(4):663–676. doi:10.1016/j.cell.2006.07.024

Takahashi K, Tanabe K, Ohnuki M, Narita M, Ichisaka T, Tomoda K, Yamanaka S (2007) Induction of pluripotent stem cells from adult human fibroblasts by defined factors. Cell 131(5):861–872. doi:10.1016/j.cell.2007.11.019

Takayama N, Nishikii H, Usui J, Tsukui H, Sawaguchi A, Hiroyama T, Eto K, Nakauchi H (2008) Generation of functional platelets from human embryonic stem cells in vitro via ES-sacs, VEGF-promoted structures that concentrate hematopoietic progenitors. Blood 111(11): 5298–5306

Tanaka T, Takahashi K, Yamane M, Tomida S, Nakamura S, Oshima K, Niwa A, Nishikomori R, Kambe N, Hara H, Mitsuyama M, Morone N, Heuser JE, Yamamoto T, Watanabe A, Sato-Otsubo A, Ogawa S, Asaka I, Heike T, Yamanaka S, Nakahata T, Saito MK (2012) Induced pluripotent stem cells from CINCA syndrome patients as a model for dissecting somatic mosaicism and drug discovery. Blood 120(6):1299–1308. doi:10.1182/blood-2012-03-417881

Thomson JA, Itskovitz-Eldor J, Shapiro SS, Waknitz MA, Swiergiel JJ, Marshall VS, Jones JM (1998) Embryonic stem cell lines derived from human blastocysts. Science 282(5391): 1145–1147

Timmermans F, Velghe I, Vanwalleghem L, De Smedt M, Van Coppernolle S, Taghon T, Moore HD, Leclercq G, Langerak AW, Kerre T, Plum J, Vandekerckhove B (2009) Generation of T cells from human embryonic stem cell-derived hematopoietic zones. J Immunol 182(11): 6879–6888

Tulpule A, Kelley JM, Lensch MW, McPherson J, Park IH, Hartung O, Nakamura T, Schlaeger TM, Shimamura A, Daley GQ (2013) Pluripotent stem cell models of Shwachman-Diamond syndrome reveal a common mechanism for pancreatic and hematopoietic dysfunction. Cell Stem Cell 12(6):727–736. doi:10.1016/j.stem.2013.04.002

Umeda K, Heike T, Yoshimoto M, Shiota M, Suemori H, Luo HY, Chui DH, Torii R, Shibuya M, Nakatsuji N, Nakahata T (2004) Development of primitive and definitive hematopoiesis from nonhuman primate embryonic stem cells in vitro. Development 131(8):1869–1879

Umeda K, Heike T, Yoshimoto M, Shinoda G, Shiota M, Suemori H, Luo HY, Chui DH, Torii R, Shibuya M, Nakatsuji N, Nakahata T (2006) Identification and characterization of hemoangiogenic progenitors during cynomolgus monkey embryonic stem cell differentiation. Stem Cells 24(5):1348–1358

Vodyanik MA, Bork JA, Thomson JA, Slukvin II (2005) Human embryonic stem cell-derived CD34+ cells: efficient production in the coculture with OP9 stromal cells and analysis of lymphohematopoietic potential. Blood 105(2):617–626

Wang Y, Jiang Y, Liu S, Sun X, Gao S (2009) Generation of induced pluripotent stem cells from human beta-thalassemia fibroblast cells. Cell Res 19(9):1120–1123

Wang Y, Zheng CG, Jiang Y, Zhang J, Chen J, Yao C, Zhao Q, Liu S, Chen K, Du J, Yang Z, Gao S (2012) Genetic correction of beta-thalassemia patient-specific iPS cells and its use in improving hemoglobin production in irradiated SCID mice. Cell Res 22(4):637–648

Warren L, Manos PD, Ahfeldt T, Loh YH, Li H, Lau F, Ebina W, Mandal PK, Smith ZD, Meissner A, Daley GQ, Brack AS, Collins JJ, Cowan C, Schlaeger TM, Rossi DJ (2010) Highly efficient reprogramming to pluripotency and directed differentiation of human cells with synthetic modified mRNA. Cell Stem Cell 7(5):618–630

Weatherall D, Clegg J (2001) The thalassaemia syndromes, 2nd edn. Wiley-Blackwell, Hoboken

Webb D, Roberts I, Vyas P (2007) Haematology of Down syndrome. Arch Dis Child Fetal Neonatal Ed 92(6):F503–F507. doi:10.1136/adc.2006.104638

Weick JP, Held DL, Bonadurer GF 3rd, Doers ME, Liu Y, Maguire C, Clark A, Knackert JA, Molinarolo K, Musser M, Yao L, Yin Y, Lu J, Zhang X, Zhang SC, Bhattacharyya A (2013) Deficits in human trisomy 21 iPSCs and neurons. Proc Natl Acad Sci U S A 110(24): 9962–9967. doi:10.1073/pnas.1216575110

Winkelstein JA, Marino MC, Johnston RBJ, Boyle J, Curnutte J, Gallin JI, Malech HL, Holland SM, Ochs H, Quie P, Buckley RH, Foster CB, Chanock SJ, Dickler H (2000) Chronic granulomatous disease. Report on a national registry of 368 patients. Medicine (Baltimore) 79(3):155–169

Woltjen K, Michael IP, Mohseni P, Desai R, Mileikovsky M, Hamalainen R, Cowling R, Wang W, Liu P, Gertsenstein M, Kaji K, Sung HK, Nagy A (2009) piggyBac transposition reprograms fibroblasts to induced pluripotent stem cells. Nature 458(7239):766–770

Yanagimachi MD, Niwa A, Tanaka T, Honda-Ozaki F, Nishimoto S, Murata Y, Yasumi T, Ito J, Tomida S, Oshima K, Asaka I, Goto H, Heike T, Nakahata T, Saito MK (2013) Robust and highly-efficient differentiation of functional monocytic cells from human pluripotent stem cells under serum- and feeder cell-free conditions. PLoS One 8(4), e59243. doi:10.1371/journal.pone.0059243

Yung SK, Tilgner K, Ledran MH, Habibollah S, Neganova I, Singhapol C, Saretzki G, Stojkovic M, Armstrong L, Przyborski S, Lako M (2013) Brief report: human pluripotent stem cell models of fanconi anemia deficiency reveal an important role for fanconi anemia proteins in cellular reprogramming and survival of hematopoietic progenitors. Stem Cells 31(5):1022–1029. doi:10.1002/stem.1308

Zou J, Mali P, Huang X, Dowey SN, Cheng L (2011a) Site-specific gene correction of a point mutation in human iPS cells derived from an adult patient with sickle cell disease. Blood 118(17):4599–4608

Zou J, Sweeney CL, Chou BK, Choi U, Pan J, Wang H, Dowey SN, Cheng L, Malech HL (2011b) Oxidase-deficient neutrophils from X-linked chronic granulomatous disease iPS cells: functional correction by zinc finger nuclease-mediated safe harbor targeting. Blood 117(21): 5561–5572

Inherited Metabolic Disorders of the Liver

7

Charis-P. Segeritz and Ludovic Vallier

Abstract

Inherited metabolic disorders (IMD) include more than 70 monogenetic disorders that are mostly associated with enzyme deficiencies. Although IMDs are rarely diagnosed, their combined incidence among full-term neonates has been estimated to account for 20 % of illnesses. One of the most common IMDs originating in the liver and necessitating transplantation is α_1-antitrypsin deficiency. Although several in vitro and in vivo models have been used to study this disease, the application of patient-specific hIPSCs and their differentiation to hepatic lineages has allowed not only physiologically relevant insights into the disease mechanisms but also provides the opportunity to understand patient-patient variabilities in phenotypes and sets the stage for the identification of novel biomarkers and the design of drugs that target them. This chapter discusses the wide application range of hIPSCs and addresses how this cellular system can meet previous shortcomings in advancing research into the molecular mechanisms of disease.

Keywords

Liver disease • Cirrhosis • hIPSC-derived hepatocytes • Alpha-1-antitrypsin deficiency

C.-P. Segeritz • L. Vallier (✉)
Wellcome Trust-MRC Stem Cell Institute, Anne McLaren Laboratory, Department of Surgery, University of Cambridge, Cambridge CB2 0SZ, UK
e-mail: lv225@cam.ac.uk

K. Fukuda (ed.), *Human iPS Cells in Disease Modelling*,
DOI 10.1007/978-4-431-55966-5_7

7.1 Introduction: Inherited Metabolic Disorders of the Liver

The liver is a unique and vital organ by the broad spectrum of its functions. It processes consumed substances and sustains reserves of iron, vitamins, and minerals. It detoxifies alcohol, drugs, and other chemicals and removes inhaled poisons such as exhaust or smoke, all of which accumulate in the bloodstream. The liver is also the site of synthesis of plasma proteins such as albumin, blood clotting factors, hormones, as well as bile required for the digestion of lipids. Finally, the liver has an essential metabolic activity by storing glycogen. Most of these activities are managed by one cell type, the hepatocyte, which constitutes the main cellular unit of the liver (Kmieć 2001). Genetic mutations targeting any gene involved in the functioning of hepatocytes can destroy the capacity of the liver to carry out these essential activities, thereby threatening the life of the afflicted patient.

Genetic diseases associated with impaired liver function are combined under the umbrella term "inherited metabolic disorders" (IMDs) which was originally coined "inborn errors of metabolism" by the British physician Archibald Garrod (1923). This class of heterogeneous diseases includes almost 70 disorders that are characterized primarily by defects in single, enzyme-coding genes that lead to lack of enzymatic activity and toxic substrate accumulation (Martins 1999). While individual events of IMDs are relatively rare, taken together they account for an incidence of 1 in 2,500–5,000 among full-term neonates (Seymour et al. 1997). Most of these disorders require medical treatment, which needs to be employed immediately after birth to avoid interference with early mental and physical development. IMDs that compromise vital liver functions are generally targeted with orthotopic liver transplantation (OLT) since only a healthy donor liver can restore the missing metabolic function. However, this intervention is associated with drawbacks since it implies high risk of surgical complications, indefinite immunosuppression associated with severe side effects, and potential organ rejection. Furthermore, the number of organ donors has remained constant during the past 10 years, while the demand for liver transplantation has more than doubled in the meantime. This situation is anticipated to worsen in the foreseeable future due to the hepatitis C pandemic and increase in cirrhosis associated with obesity. The development of alternative therapies to OLT has therefore become a major objective in the field of regenerative medicine. Further goals for the translational research community include the advancement of physiologically relevant cell models to study the underlying disease mechanisms of the respective disorders and the design and screening of novel drug compounds.

7.2 An Example: α_1-Antitrypsin Deficiency

α_1-antitrypsin (A1AT) deficiency is one of the most common IMDs, affecting 1 in 2000 individuals of North Caucasian decent (Sveger 1976). A1AT deficiency is characterized by a point mutation in the A1AT gene (*SERPINA1*), which is primarily expressed and synthesized by hepatocytes and secreted as a circulating plasma glycoprotein. While wild-type A1AT acts as an antiprotease that enzymatically inhibits overactive neutrophil elastase in pulmonary tissues, defective A1AT protein synthesis results in intracellular A1AT protein polymerization, accumulation and reduced A1AT levels in the circulation and in target organ tissues such as the lungs. A1AT deficiency is inherited as an autosomal codominant trait leading to homozygous or heterozygous phenotypes. Homozygous A1AT-deficient individuals are at an increased risk of developing chronic hepatitis, liver cirrhosis, and chronic-obstructive respiratory disease. This chapter introduces the pathophysiology of A1AT deficiency and explains the suitability of human-induced pluripotent stem cells (hIPSCs) to study this disease and unearth unknown mechanisms involved in its pathogenesis.

7.2.1 α_1-Antitrypsin

A1AT is a 52 kDa acute phase glycoprotein synthesized mainly in the liver but at low levels is also locally produced by blood monocytes, tissue macrophages, and other non-hepatic tissues, such as airway epithelial cells (Carlson et al. 1988; Perlmutter et al. 1985). Secreted into the blood stream, A1AT functions as the most abundant circulating protease inhibitor in man, acting in all organ tissues but most prominently in the lung (Morrison et al. 1984). As member of the serine protease inhibitor (serpin) superfamily, A1AT serves to prevent pulmonary tissue damage from overactive proteases, particularly human neutrophil elastase (Travis and Salvesen 1983). Neutrophil elastases are active in several inflammatory and degenerative diseases (Jenne 1994). They are stored alongside other serine proteases (e.g., cathepsin G) in azurophil granules of neutrophils and cause hydrolytic, proteolytic, and oxidative breakdown of phagocytosed compounds (e.g., bacteria and viruses) (Sandborg and Smolen 1988). However, lysosomal enzymes may escape from activated neutrophils and digest connective tissue matrix proteins (e.g., elastin) (Janoff and Scherer 1968). To prevent this, the pulmonary environment is generally bathed in excess A1AT and supports A1AT molecules to form specific complexes with free proteases. This inactivates the enzyme through conformational deformation and is achieved through the spatial arrangement of A1AT's single β-sheet A, nine α-helices, and a mobile reactive-center loop peptide that all create a pseudo-substrate for neutrophil elastase (Eriksson 1996; Huntington et al. 2000). Following binding, the cleaved reactive-center loop peptide inactivates bound neutrophil elastase by inserting into A1AT's central β-sheet A (Huntington et al. 2000) (Fig. 7.1, left panel). In the event of imbalances between neutrophil elastase and A1AT levels, as observed in A1AT deficiency, the reduction of inhibitory activity from A1AT can lead to protease-induced destruction of pulmonary connective tissue (Janoff 1985) which may manifest as emphysema (Tobin et al. 1983).

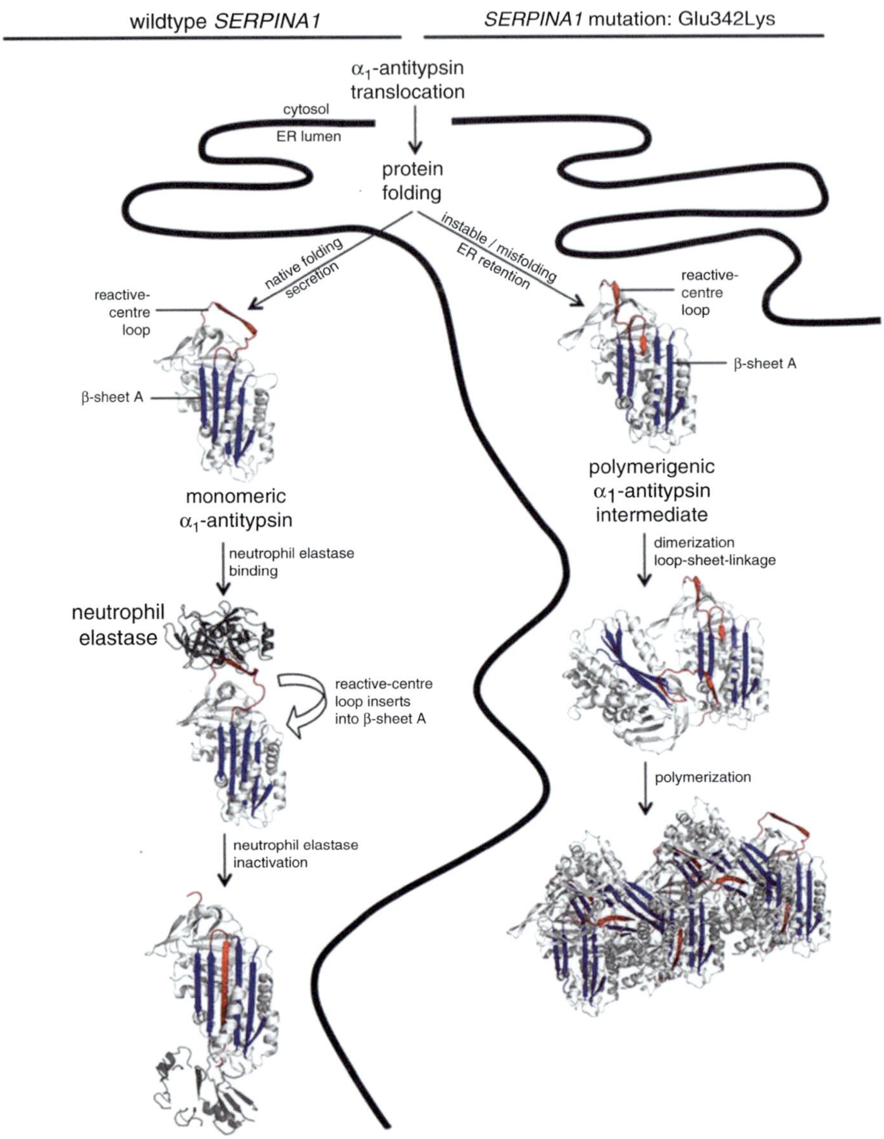

Fig. 7.1 Conformational changes and differences in function between wild-type and mutant α_1-antitrypsin protein variants. Wild-type *SERPINA1* sequences result in native protein folding of α_1-antitrypsin (A1AT) protein and secretion into the circulation where upon A1AT's reactive-center loop binding to neutrophil elastase, it is cleaved and inserted into A1AT's β-sheet A, resulting in enzymatic inactivation of neutrophil elastase. Mutations in *SERPINA1* undermine this process and lead to instable A1AT intermediates that readily polymerize (Structural protein images were reproduced with permission from Dr. Bibek Gooptu (Gooptu and Lomas 2008))

7.2.2 Genetic Basis, Molecular Mechanisms, and Epidemiology

Although more than 120 allelic variants of A1AT have been discovered to date, only a minority of those have been linked with pathological A1AT deficiency (Brantly et al. 1988; DeMeo and Silverman 2004). The most prevalent A1AT variants in white ethic populations are PiM (~95 %), PiS (2–3 %), and PiZ (1–3 %) alleles (Luisetti and Seersholm 2004). While the M allele annotates wild-type A1AT, the S and Z allele signify deficient A1AT variants caused by point mutations in *SERPINA1* (Jeppsson 1976; Owen and Carrell 1976). The most common and severe form of clinically diagnosed A1AT deficiency is linked to the Z allele, which is dependent on a point mutation-induced amino acid substitution of a lysine for a glutamic acid residue in *SERPINA1*'s exon 5 (Glu342Lys) (Jeppsson 1976). The most supported mechanism for the polymerization of its protein product has been linked to mutant Z A1AT's conformational instability and termed "loop-sheet" polymerization (Lomas et al. 1992). It describes the widening of A1AT's β-sheet A as a result of the amino acid substitution at position 342 and the subsequent insertion of the reactive-center loop of another A1AT polypeptide into the opened β-sheet A, resulting in irreversible linkage of two Z A1AT molecules (Lomas et al. 1992) (Fig. 7.1, right panel). While the M A1AT allele contributes to 50 % of the overall serum A1AT levels, the Z allele supplies only ~5 % to circulating A1AT plasma concentrations (Fregonese and Stolk 2008). As a result, homozygous PiZZ patients are characterized by plasma A1AT deficiency levels of ~10–15 %. The remaining synthesized A1AT is prevented from secretion through intracellular degradation (~70 %) and accumulation (~15 %) of polymers that can be detected as polymer filaments with electron microscopy or as intrahepatic endoplasmic reticulum (ER) inclusions (Lomas et al. 1992) (Fig. 7.2). The Z allele is most prevalent in Northern European countries, where 0.10 % are affected by Z-related-deficient allele combinations (PiSZ, PiZZ) (Luisetti and Seersholm 2004; Sveger 1976; Serres 2002). Worldwide there are 3.4 million individuals with pathological allele deficiencies (PiSS, PiSZ, PiZZ) (Serres 2002; American Thoracic Society and European Respiratory 2003).

7.2.3 Clinical Symptoms

The significant reduction of secreted A1AT is caused by pathogenic accumulation of A1AT in the liver of homozygous PiZZ individuals (Laurell and Eriksson 1963). This leads to toxic gain-of-function mechanisms in hepatocytes and marks A1AT deficiency as the most common genetic liver disease affecting children and requiring juvenile liver transplantation (Gartner Jr et al. 1984). A1AT deficiency can present as neonatal cholestasis (jaundice), juvenile hepatitis, enlarged spleen, and swelling of the abdomen during the first months or years of life. 10–15 % of A1AT-deficient children and 40–60 % of overall PiZZ patients will develop chronic liver disease at some stage, which may advance to liver cirrhosis or liver cancer (Sveger 1976, 1988; Eriksson et al. 1986).

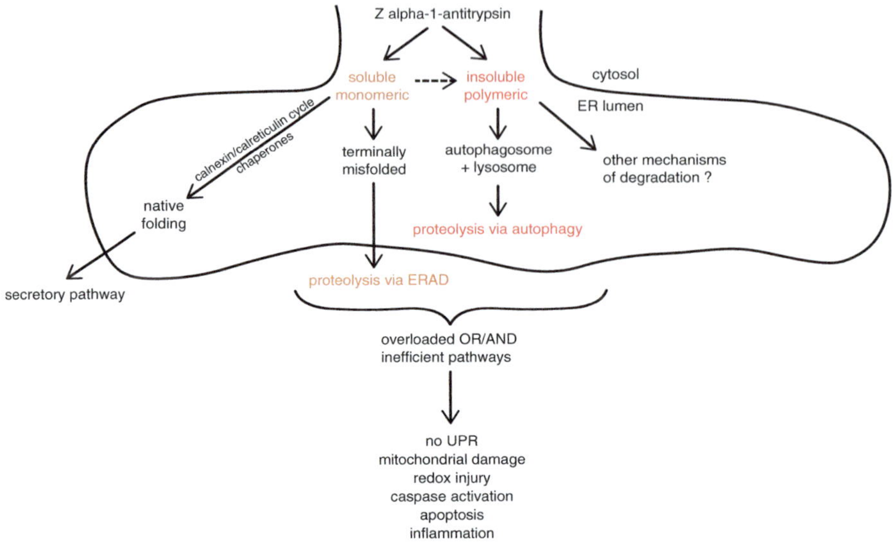

Fig. 7.2 Intracellular processing of mutant α_1-antitrypsin protein. The synthesis and translocation of mutant Z α_1-antitrypsin (A1AT) protein into the endoplasmic reticulum (ER) lumen results in formation of soluble monomeric protein units, which may further assemble to form insoluble polymer aggregates. With the help of chaperones, a minority of mutant A1AT monomers attain native folding and will be processed in the Golgi apparatus for secretion. Terminally misfolded A1AT proteins are bound by BiP and retro-translocated from the ER to the cytosol for subsequent ubiquitination and proteasomal degradation (ERAD). Insoluble A1AT polymer aggregates are removed from hepatic ER through autophagy, which eliminates protein aggregates, damaged ER, and other organelles via lysosomal degradation pathways. Due to the continued synthesis of mutant Z A1AT, these disposal pathways can become overwhelmed and fail to resolve ER stress inferred by protein accumulation. As a result, affected cells display signs of inflammation and intracellular injury and may trigger apoptotic pathways

Similar to lung disease, there is great variability between A1AT-deficient patients in the phenotypic expression of associated liver disease. The onset, progression, and severity of liver disease have been shown to vary widely even between PiZZ siblings (Hinds et al. 2006). Generally, higher rates of degradation of misfolded intracellular A1AT polymers have been linked with reduced liver disease burden (Wu et al. 1994). As such, mechanisms to upregulate autophagy have been tied to reduced A1AT polymer concentrations and less liver injury in a mouse model (Kaushal et al. 2010). To date, the only curative intervention for end-stage A1AT deficiency is whole-organ liver transplantation (Fairbanks and Tavill 2008). The overall mortality of individuals with severe A1AT deficiency has been examined in several studies and has been reported between 18 and 37 % (Larsson 1978; Seersholm et al. 1994; Stoller et al. 2005).

Until recently, the detailed molecular mechanisms causing Z A1AT polymer accumulation in the ER lumen of hepatocytes to trigger liver disease were not well understood. Improved insights into these processes would improve the understanding of what distinguishes the A1AT-deficient subpopulation susceptible to liver disease (~20 %) from those protected (~80 %) (Fairbanks and Tavill 2008). Given that the mere Glu342Lys amino acid substitution in *SERPINA1* cannot account for the emergence of liver disease, PiZZ individuals without liver disease likely carry certain genetic traits that protect from A1AT polymer-associated disease. Alternatively, PiZZ individuals with liver disease may be exposed to environmental factors or carry a specific genetic deposition that renders them susceptible to disease. Several groups have attributed the buildup of A1AT aggregates and resultant liver injury to impaired and delayed degradation of Z A1AT (Wu et al. 1994). This may be due to compromised interactions between Z A1AT and factors of the calnexin/calreticulin cycle, which generally aid glycoproteins in their folding and tag permanently misfolded glycoproteins for proteasomal degradation (Cameron et al. 2009; Qu et al. 1996). Besides impaired proteasomal degradation of Z A1AT molecules, elimination of A1AT polymers via autophagy may also be reduced, delayed, or overwhelmed (Perlmutter 2011). Other studies have shown that A1AT polymer accumulation activates the NF-kB pathway independent of the unfolded protein response (UPR), triggering toxic ER stress signals in protein-overloaded hepatocytes (Hidvegi et al. 2005; Ordóñez et al. 2013). The absence of the UPR may prevent downstream pathways from eliminating misfolded proteins and appropriately regulate apoptosis under chronic ER stress (Hidvegi et al. 2005). It also has been proposed that A1AT polymer-bearing hepatocytes cannot recognize the ordered structures of A1AT polymers as misfolded proteins and therefore fail to mount strong protein disposal pathways (Graham et al. 1990; Hidvegi et al. 2005).

7.3 In Vitro and In Vivo Models: How to Study α_1-Antitrypsin Deficiency

As outlined in the previous sections, A1AT deficiency is a liver disease with substantial health consequences but limited options for treatment. In the past, the pressing need to gain insights into the underlying disease mechanisms and the development of new therapeutic tools has been hampered by limited availability and viability of primary A1AT-deficient and wild-type hepatocytes to study the disease. Although animal models have contributed to improved understanding of A1AT deficiency, physiological variability between animal and man has questioned the extrapolation of findings into a *Homo sapiens* context. The advent of human-induced pluripotent stem cells (hIPSCs) derived from patients with A1AT deficiency and their differentiation to the hepatic lineage has generated cells which phenocopy the disease phenotypes and can target these aforementioned obstacles (Rashid et al. 2010). This section discusses the use and shortcomings of alternative A1AT deficiency models and how hIPSC-derived hepatocytes have developed into a suitable platform for investigating the pathophysiology of A1AT deficiency.

7.3.1 Primary PiZZ Hepatocytes

Primary hepatocytes isolated from PiZZ patients are the most appropriate cell model to study A1AT deficiency. However, it requires invasive biopsies to access the liver tissue. Furthermore, multiple groups have failed to extract and plate high-quality PiZZ hepatocytes from cirrhotic liver tissue which can become available during liver transplantations (Gramignoli et al. 2013). At this stage of disease severity, PiZZ hepatocytes are generally incorporated into a highly fibrotic tissue and are therefore very difficult to isolate, let alone culture (Bhogal et al. 2011). At earlier stages of disease when the liver is still pre-fibrotic, isolation of PiZZ hepatocytes through enzymatic digestion of liver explant tissue may be more feasible. However, oftentimes PiZZ patients are asymptomatic and therefore remain undiagnosed at that point, making it difficult to obtain biological material. Furthermore, due to the absence of long-term culture systems for primary hepatocytes, the study of disease progression over time cannot be accomplished. Finally, the dissociation of hepatocytes from their niche liver microenvironment and subsequent plating in 2D conditions is well known to cause loss of hepatic functionality in primary hepatocytes (Bhatia et al. 2014; Gramignoli et al. 2013).

7.3.2 Evaluation of Alternative Models to Study α_1-Antitrypsin Deficiency

In the past, several in vitro and in vivo models have aided in advancing the A1AT deficiency field. For example, cloning of human PiM A1AT into the murine germ line generated transgenic mice that expressed human A1AT levels in serum and liver (Carlson et al. 1988; Sifers et al. 1987). Using the same technique, Carlson et al. and others generated transgenic mice expressing the human PiZ A1AT gene product. These animals exhibited mutant A1AT synthesis and showed retention of mutant A1AT in hepatic ER, hepatocellular damage, and chronic liver injury that correlated with the level of hepatic PiZ A1AT accumulation. The phenotypes in pups resembled human neonatal hepatitis (Carlson et al. 1989; Dycaico et al. 1988). Despite the significant increase in inflammation recorded in these transgenic PiZ mice, there was no statistically significant increase in fibrosis. This was attributed to the faster metabolism and shorter life expectancy of mouse versus man, which may not suffice for the development of chronic disease, including fibrosis, cirrhosis, and hepatocellular carcinoma.

In order to evaluate the pathogenic impact of an elastase to antielastase imbalance on pulmonary health, the pallid mouse has been used to study A1AT deficiency (Martorana et al. 1993). This mouse strain suffers from hereditary emphysema and develops pulmonary lesions comparable with human emphysema. While the hepatic A1AT activity is not disrupted in this mouse model, circulating A1AT levels are reduced and exhibit a diminished capacity for elastase inhibition. Despite the suitability of studying the pallid mouse for its reduced serum elastase inhibitory effects and resultant lung tissue damage, it cannot provide insight into the underlying mechanisms of hepatic disease in PiZZ A1AT humans (Martorana et al. 1993).

Several other yeast and mammalian cell line models have also been shown to recapitulate features of A1AT deficiency (Gosai et al. 2010; Wu et al. 1994; Ordóñez et al. 2013; Cabezn et al. 1984; Kruse et al. 2006; McCracken and Brodsky 1996; Scott et al. 2007; Werner et al. 1996). Despite the exciting findings obtained from these studies, the lack of physiological human A1AT expression makes it difficult to draw conclusions into the pathogenesis of A1AT deficiency. Multiple A1AT gene copy numbers, interference of endogenous animal antiproteases, a genetic background that may not permit onset of disease upon A1AT polymer accumulation, and the absence of endogenous promoters and transcription factors to activate A1AT expression all contribute to a likely non-physiological system to study A1AT deficiency. Furthermore, differences in tissue composition, anatomy, physiology, lack of conservation of gene order, and the animal's inbred genetic background can direct *SERPINA1* mutations to phenotypes that may not faithfully phenocopy the clinical picture of human A1AT deficiency (Shapiro 2007). For instance, instead of a single gene, as observed in primates, murine *SERPINA1* exists as a small gene family of up to five genes that synthesize several isoforms (Borriello and Krauter 1991). It still remains to be determined which member/s of this gene cluster correspond to the functional equivalent of human A1AT (Barbour et al. 2002).

7.4 Modeling α_1-Antitrypsin Deficiency Using Human-Induced Pluripotent Stem Cells

7.4.1 Human-Induced Pluripotent Stem Cells

hIPSCs, derived from patients with specific disease backgrounds and their subsequent differentiation to specialized somatic cells in which the disease manifests, have emerged as in vitro systems to model various diseases ex vivo. The derivation of hIPSCs followed shortly after the very first induced pluripotent stem cell lines were generated from mouse embryonic fibroblasts (Takahashi et al. 2007; Takahashi and Yamanaka 2006). This was achieved through forced expression of transcription factors c-Myc, Oct3/4, Sox2, and Klf-4 (Yamanaka factors) in fibroblasts. While this process of direct reprogramming was previously performed with viral vectors, several techniques are now available that use non-integrative methods to safely deliver the reprogramming factors (e.g., episomal vectors) (Yu et al. 2011). hIPSCs resemble embryonic stem cells (pluripotent stem cells derived from an embryo at the blastocyst stage) in terms of their expression of pluripotent genes, DNA methylation patterns, doubling time, formation of embryoid bodies, teratoma formation, and viable chimeras, as well as their potency for self-renewal and differentiation into almost any cell type (Wernig et al. 2007). hIPSC research and the development of differentiation protocols to generate specific cell types have promised to yield great advancements for the disease modeling field but also for developmental biology, personalized cell-based therapy, and the generation of platforms to develop and screen new pharmaceuticals (Rashid and Vallier 2010). In the field of transplantation and regenerative medicine, the possibility to produce and implant personalized hIPSC-derived somatic cells into patients removes the risk of tissue rejection inherent to allogeneic transplants, high costs associated with invasive transplantation, and lifelong administration of immunosuppressive drugs and bypasses some of the ethical issues associated with embryonic stem cells.

7.4.2 Derivation and Hepatic Differentiation of α_1-Antitrypsin-Deficient hIPSC

In the search for human cellular systems that closely mimic human A1AT-deficient hepatocyte dysfunction, our group and others have generated hIPSC lines from PiZZ patients and developed hepatic differentiation protocols to direct hIPSCs along their in vivo developmental path to form hepatocyte-like cells (A1AT-ZZ) (Rashid et al. 2010; Tafaleng et al. 2015; Wilson et al. 2015) (Fig. 7.3). This was achieved by culturing hIPSCs in changing cytokine combinations and concentrations (e.g., FGF2, activin A, BMP4, Wnt, OSM, and HGF), as well as media compositions that mimic the microenvironment during hepatic embryogenesis in vivo (Si-Tayeb et al. 2010; Touboul et al. 2010). This process includes the transition from pluripotency to definitive endoderm and its subsequent anteriorization (foregut); specification into hepatic endoderm, which develops cells into bipotential hepatocytes; and finally hepatic maturation to generate hepatocytes. The classification of hIPSC-derived cells as "hepatocytes" (or "hepatocyte-like cells") is based on their ability to recapitulate key features of

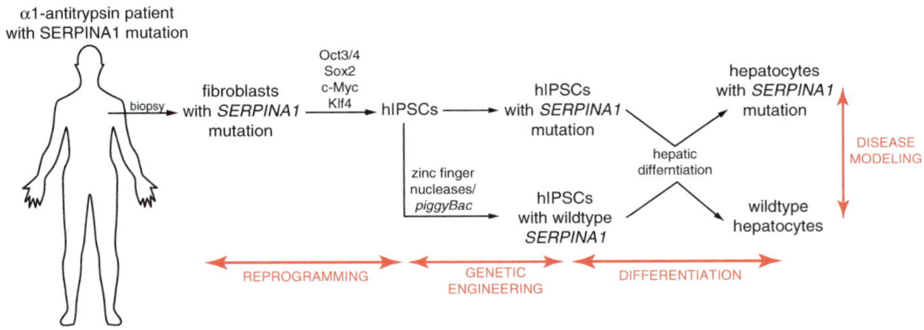

Fig. 7.3 Modeling α_1-antitrypsin deficiency using hIPSC-derived hepatocytes. Fibroblasts from a PiZZ α_1-antitrypsin-deficient individual were reprogrammed to generate human-induced pluripotent stem cells (hIPSCs). While this cell line carried the disease-specific *SERPINA1* mutation, corrective genome-editing techniques resulted in an isogenic wild-type hIPSC line. Both cell lines differentiated to hepatic lineage allow for comparative disease modeling of α_1-antitrypsin deficiency

their in vivo equivalents. This includes co-expression of hepatic genes (e.g., HNF4α, A1AT), hepatic protein synthesis, and secretion (e.g., transthyretin, albumin), as well as liver-specific functionalities such as metabolic enzyme activity (cytochrome P450) and liver engraftment in animal models of acute liver failure (Rashid et al. 2010; Hengstler et al. 2005; Schwartz et al. 2014). Importantly, hepatocyte-like cells generally still feature low expression of hepatic maturity markers when compared to primary hepatocytes (Sampaziotis et al. 2015). Since low maturity of hIPSC-derived hepatocytes is a shared shortcoming of other differentiation protocols, many groups are currently focused on advancing this field (e.g., Huang et al. 2009).

In addition, A1AT-ZZs exhibited all disease-specific characteristics of A1AT deficiency, including A1AT polymer formation, intracellular retention in the hepatic ER, and reduced A1AT secretion (Rashid et al. 2010; Tafaleng et al. 2015; Wilson et al. 2015). Thus, A1AT-ZZ hepatocyte-like cells represent an attractive system to study the molecular mechanisms directing the disease pathophysiology.

7.5 Applications of α_1-Antitrypsin-Deficient hIPSC-Derived Hepatocytes

7.5.1 Cell Therapy: Genetic Correction of Diseased hIPSCs to Generate Wild-Type hIPSC

In order to generate a wild-type cell line, our group employed zinc finger nucleases and *piggyBac*-based genome-editing techniques to correct the *SERPINA1* mutation in hIPSCs responsible for A1AT deficiency, creating an otherwise isogenic wild-type hIPSC line (Yusa et al. 2011). This cell line was termed A1AT-RR, referring to the corrected *SERPINA1* gene. Once differentiated to hepatic lineages, the disease phenotype was rescued in this cell type and A1AT-RRs exhibited normal levels of A1AT secretion as well as the absence of intracellular A1AT polymer formation or accumulation (Yusa et al. 2011). The resulting wild-type hIPSC-derived hepatocytes were transplanted successfully into animal models for liver failure, providing a first proof of principle that hIPSCs could be used for personalized medicine of inherited metabolic diseases (Yusa et al. 2011). With increased hepatic maturity, we anticipate that transplantation of these cells will be able to correct hepatic functionality.

7.5.2 Disease Modeling: Linking Hepatic Accumulation of α1-Antitrypsin Polymers to Pathological Liver Conditions

Despite the volume of information available about the intracellular polymerization of A1AT, it has been unclear how the A1AT polymer retention leads to liver disease. Due to the capacity of A1AT-deficient hIPSC-derived hepatocytes to be cultured long-term, it is possible to observe the accumulation of A1AT polymers over time. Studies aimed at investigating the functional divergences between diseased and healthy hepatocytes can utilize A1AT-ZZs and A1AT-RRs, which only differ in the absence or presence of the *SERPINA1* mutation. This allows for detailed comparisons of genetic and protein material at modeled end-stage liver disease, circumventing the interference of differing genetic backgrounds. Using this cellular platform, we have been able to dissect potential pathways involved in linking polymer accumulation and liver disease and allowed us to detect novel biomarkers for this disease (Segeritz et al. in preparation). Interestingly, recent genome-wide studies have also revealed novel aspects of the disease including UPR activity early during the process of hepatocyte specification (Wilson et al. 2015). Thus, these data which would have been impossible to obtain in any other model system suggest that liver disease associated with A1AT could be induced during early liver development.

7.5.3 Patient-Patient Variability: Differentiating the Susceptible from the Protected

hIPSC-derived hepatocytes derived from A1AT-deficient patients have also aided in addressing the question of why some affected individuals are protected from acquiring liver disease, while others suffer from cirrhotic livers that require liver transplantation. The derivation of hIPSCs from A1AT-deficient patients with and without disease and their differentiation to hepatocyte-like cells has allowed insights into the individual disease phenotypes (Tafaleng et al. 2015). While the intracellular Z A1AT in hIPSC-derived hepatocytes of patients without liver disease was primarily degraded by autophagic pathways, elimination of Z A1AT was delayed in hIPSC-derived hepatocytes of patients that suffered from liver disease. Z A1AT in the latter group also appeared to accumulate in ER inclusions. This work suggested that the observed variation in liver disease phenotypes of A1AT deficiency is linked to the rate of misfolded A1AT degradation. This opens the door for the development and testing of specific drug targets that control proteostasis and enhance intracellular degradation such as carbamazepine (Hidvegi et al. 2010). Furthermore, these observations also indicate that hIPSC-derived hepatocytes can be used to predict disease progression and susceptibility, allowing for the implementation of personalized therapeutics.

7.5.4 Drug Development and Drug Screening: Finding a Cure for α1-Antitrypsin Deficiency

Due to the current absence of therapies for the treatment of A1AT deficiency, there is great urgency to develop and test new drug compounds in human systems. hIPSCs derived from A1AT-deficient individuals and differentiated to hepatocytes could act as a suitable platform to screen efficacy of novel drug compounds. Previously, hIPSC-derived hepatocytes have been employed in drug toxicity studies and to show patient-specific sensitivity to a given drug compound (Cayo et al. 2012; Takayama et al. 2013). However, for the first time, a recent separate study has shown that carbamazepine could improve autophagic flux in A1AT-ZZ hepatocyte-like cells (Wilson et al. 2015), thereby reinforcing the interest of clinical trial currently testing this small molecule for A1AT deficiency. Considered together these results exemplify how patient-specific, hIPSC-derived hepatocytes could be used to screen drug compounds to reverse or prevent the accumulation of A1AT polymers.

7.6 Conclusion

Using the example of A1AT deficiency as a representative IMD, this chapter has demonstrated the suitability of patient-specific hIPSC to model metabolic disorders affecting the liver. Indeed, hIPSCs derived from patient with specific IMDs can be easily differentiated into somatic cell types in which the disease manifests to address questions pertinent to the disease mechanism, the variability in disease phenotypes, and the development of treatment regimens. Thus, hIPSC-based hepatocyte platforms represent a physiologically relevant system to gain new understandings of IMDs and design strategies to reduce the burden of these diseases.

References

American Thoracic Society and European Respiratory (2003) American Thoracic Society Documents American Thoracic Society/European Respiratory Society Statement: standards for the diagnosis and management of individuals with. Am J Respir Clin Care Med 168: 818–900. doi:10.1164/rccm.168.7.818

Barbour KW, Wei F, Brannan C, Flotte TR, Baumann H, Berger FG (2002) The murine alpha-1-proteinase inhibitor gene family: polymorphism, chromosomal location, and structure. Genomics 80(5):515–522. doi:10.1006/geno.2002.6864

Bhatia SN, Underhill GH, Zaret KS, Fox IJ (2014) Cell and tissue engineering for liver disease. Sci Transl Med 6(245):1–21

Bhogal RH, Hodson J, Bartlett DC, Weston CJ, Curbishley SM, Haughton E, Williams KT et al (2011) Isolation of primary human hepatocytes from normal and diseased liver tissue: a one hundred liver experience. PLoS One 6(3):1–8. doi:10.1371/journal.pone.0018222

Borriello F, Krauter KS (1991) Multiple murine alpha-1-protease inhibitor genes show unusual evolutionary divergence. Proc Natl Acad Sci 88:9417–9421

Brantly M, Nukiwa T, Crystal RG (1988) Molecular basis of alpha-1-antitrypsin deficiency. Am J Med 84(6A):13–31

Cabezn T, Wilde MDE, Heriont P, Loriaut R, Bollent A (1984) Expression of human a1-antitrypsin cDNA in the yeast Saccharomyces cerevisiae. Proc Natl Acad Sci U S A 81: 6594–6598

Cameron PH, Chevet E, Thomas DY, John JM, Cameron PH, Chevet E, Pluquet O et al (2009) Calnexin phosphorylation attenuates the release of partially misfolded alpha-1-antitrypsin to the secretory pathway. J Biol Chem 284:34570–34579. doi:10.1074/jbc.M109.053165

Carlson JA, Rogers BB, Sifers RN, Hawkins HK, Finegold MJ, Woo SL (1988) Multiple tissues express alpha 1-antitrypsin in transgenic mice and man. J Clin Invest 82(1):26–36
Carlson JA, Rogers BB, Sifers RN, Finegold MJ, Clift SM, DeMayo FJ, Bullock DW et al (1989) Accumulation of iZ alpha 1-antitrypsin causes liver damage in transgenic mice. J Clin Invest 83(4):1183–1190
Cayo MA, Cai J, Delaforest A, Noto FK, Nagaoka M, Clark BS, Collery RF et al (2012) JD induced pluripotent stem cell–derived hepatocytes faithfully recapitulate the pathophysiology of familial hypercholesterolemia. Hepatology 56(6):2163–2171. doi:10.1002/hep.25871
DeMeo DL, Silverman EK (2004) Alpha-1-antitrypsin deficiency. 2: genetic aspects of alpha-1-antitrypsin deficiency: phenotypes and genetic modifiers of emphysema risk. Thorax 59: 259–264. doi:10.1136/thx.2003.006502
Dycaico MJ, Grant SGN, Felts K, Nichols WS, Geller SA, Hager JH, Pollard AMYJ et al (1988) Neonatal hepatitis induced by alpha-1-antitrypsin: a transgenic mouse model. Science 242: 1409–1412
Eriksson S (1996) A 30-year perspective on alpha-1-antitrypsin deficiency. Chest 110(6):237–242
Eriksson S, Carlson J, Velez R (1986) Risk of cirrhosis and primary liver cancer in alpha-1-antitrypsin deficiency. N Engl J Med 314:736–739
Fairbanks KD, Tavill AS (2008) Liver disease in alpha 1-antitrypsin deficiency: a review. Am J Gastroenterol 103(8):2136–2141. doi:10.1111/j.1572-0241.2008.01955.x; quiz 2142
Fregonese L, Stolk J (2008) Hereditary alpha-1-antitrypsin deficiency and its clinical consequences. Orphanet J Rare Dis 3(16):1–9. doi:10.1186/1750-1172-3-16
Garrod AE (1923) Inborn errors of metabolism, vol 2. Henry Frowde and Hodder & Stoughton, London
Gartner JC Jr, Zitelli BJ, Malatack JJ, Shaw BW, Iwatsuki S, Starzl TE (1984) Orthotopic liver transplantation in children: two-year experience with 47 patients. Pediatrics 74(1):140–145
Gooptu B, Lomas DA (2008) Polymers and inflammation: disease mechanisms of the serpinopathies. J Exp Med 205(7):1529–1534. doi:10.1084/jem.20072080
Gosai SJ, Kwak JH, Luke CJ, Long OS, King DE, Kovatch KJ, Paul A, et al (2010) Automated high-content live animal drug screening using C. elegans expressing the aggregation prone serpin a1-antitrypsin Z. PLoS One 5(11):e15460. doi:10.1371/journal.pone.0015460
Graham K, Le A, Sifers R (1990) Accumulation of the insoluble PiZ variant of human alpha 1-antitrypsin within the hepatic endoplasmic reticulum does not elevate the steady-state level of grp78/BiP. J Biol Chem 265:20463–20468
Gramignoli R, Tahan V, Dorko K, Skvorak KJ, Hansel MC, Zhao W, Venkataramanan R et al (2013) New potential cell source for hepatocyte transplantation: discarded livers from metabolic disease liver transplants. Stem Cell Res 11(1):563–573. doi:10.1016/j.scr.2013.03.002.New
Hengstler J, Brulport M, Schormann W, Bauer A, Hermes M, Nussler A, Fandrich F et al (2005) Generation of human hepatocytes by stem cell technology: definition of the hepatocyte. Expert Opin Drug Metab Toxicol 1(1):61–74
Hidvegi T, Schmidt BZ, Hale P, Perlmutter DH, Hidvegi T, Schmidt BZ, Hale P et al (2005) Accumulation of mutant alpha1-antitrypsin Z in the endoplasmic reticulum activates caspases-4 and -12, NF-KB, and BAP31 but not the unfolded protein response. J Biol Chem 280: 39002–39015. doi:10.1074/jbc.M508652200
Hidvegi T, Ewing M, Hale P, Dippold C, Beckett C, Kemp C, Maurice N et al (2010) An autophagy-enhancing drug promotes degradation of mutant alpha-1-antitrypsin Z and reduces hepatic fibrosis. Science 329:229–232. doi:10.1126/science.1190354
Hinds R, Hadchouel A, Shanmugham NP, Al-Hussaini A, Chambers S, Cheeseman P, Mieli-Vergani G et al (2006) Variable degree of liver involvement in siblings with PiZZ alpha-1-antitrypsin deficiency-related liver disease. J Pediatr Gastroenterol Nutr 43:136–138
Huang P, Zhang L, Gao Y, He Z, Yao D, Wu Z, Cen J et al (2009) Article direct reprogramming of human fibroblasts to functional and expandable hepatocytes. Stem Cell 14(3):370–384. doi:10.1016/j.stem.2014.01.003. Elsevier Inc

Huntington JA, Read RJ, Carrell RW (2000) Structure of a serpin-protease complex shows inhibition by deformation. Nature 407:923–926

Janoff A (1985) Elastases and emphysema. Current assessment of the protease-antiprotease hypothesis. Am Rev Respir Dis 132(2):417–433

Janoff A, Scherer J (1968) Mediators of inflammation in leukocyte lysosomes – IX. Elastinolytic activity in granules of human polymorphonuclear leukocytes. J Exp Med 128(5):1137–1155

Jenne DE (1994) Structure of the azurocidin, proteinase 3, and neutrophil elastase genes – implications for inflammation and vasculitis. Am J Respir Crit Care Med 150:S147–S154

Jeppsson JO (1976) Amino acid substitution Glu leads to Lys alpha1-antitrypsin PiZ. FEBS Lett 65(2):195–197

Kaushal S, Annamali M, Blomenkamp K, Rudnick D, Halloran D, Brunt E, Teckman J (2010) Rapamycin reduces intrahepatic alpha-1-antitrypsin mutant Z protein polymers and liver injury in a mouse model. Exp Biol Med 235(6):700–709. doi:10.1258/ebm.2010.009297.Rapamycin

Kmieć Z (2001) Cooperation of liver cells in health and disease. Adv Anat Embryol Cell Biol 161(III–XIII):1–151

Kruse KB, Brodsky JL, McCracken AA (2006) Characterization of an ERAD gene as VPS30/ATG6 reveals two alternative and functionally distinct protein quality control pathways: one for soluble Z variant of human alpha-1 proteinase inhibitor (A1PiZ) and another for aggregates of A1PiZ. Mol Biol Cell 17:203–212. doi:10.1091/mbc.E04

Larsson C (1978) Natural history and life expectancy in severe alpha-1-antitrypsin deficiency, PiZ. Acta Med Scand 204(5):345–351

Laurell CB, Eriksson S (1963) The electrophoretic α 1 -globulin pattern of serum in α 1 -antitrypsin deficiency. Scand J Clin Lab Invest 15(2):132–140. doi:10.3109/15412555.2013.771956

Lomas DA, Li-Evans D, Finch JT, Carrell RW (1992) The mechanisms of Z alpha-1-antitrypsin accumulation in the liver. Nature 357:605–607

Luisetti M, Seersholm N (2004) Alpha-1-antitrypsin deficiency – 1: epidemiology of alpha-1-antitrypsin deficiency. Thorax 59:164–169

Martins A (1999) Inborn errors of metabolism: a clinical overview. Sao Paulo Med J 117(6):251–265

Martorana PA, Brand T, Gardi C, van Even P, de Santi MM, Calzoni P, Marcolongo P, Lungarella G (1993) The pallid mouse. A model of genetic alpha 1-antitrypsin deficiency. Lab Invest 68(2):233–241. Lab Invest

McCracken AA, Brodsky JL (1996) Assembly of ER-associated protein degradation in vitro: dependence on cytosol, calnexin, and ATP. J Cell Biol 132(3):291–298

Morrison HM, Afford SC, Stockley RA (1984) Inhibitory capacity of alpha-1-antitrypsin in lung secretions: variability and the effect of drugs. Thorax 39:510–516

Ordóñez A, Snapp EL, Tan L, Miranda E, Marciniak SJ, Lomas DA (2013) Endoplasmic reticulum polymers impair luminal protein mobility and sensitise to cellular stress in α1-antitrypsin deficiency. Hepatology 57(5):2049–2060. doi:10.1002/hep.26173.Endoplasmic

Owen MC, Carrell RW (1976) Alpha-1-antitrypsin: molecular abnormality of S variant. Br Med J 1:130–131

Perlmutter DH (2011) Alpha-1-antitrypsin deficiency: importance of proteasomal and autophagic degradative pathways in disposal of liver disease–associated protein aggregates. Annu Rev Med 62:333–345. doi:10.1146/annurev-med-042409-151920

Perlmutter DH, Cole FS, Kilbridge P, Rossing T (1985) Expression of the alpha-1-proteinase inhibitor gene in human monocytes and macrophages. Proc Natl Acad Sci 82:795–799

Qu D, Teckman JH, Omura S, Perlmutter DH (1996) Degradation of a mutant secretory protein, α-1-antitrypsin Z, in the endoplasmic reticulum requires proteasome activity. J Biol Chem 271:22791–22795. doi:10.1074/jbc.271.37.22791

Rashid ST, Vallier L (2010) Induced pluripotent stem cells – alchemist's tale or clinical reality? Expert Rev Mol Med 12:25

Rashid ST, Corbineau S, Hannan N, Marciniak SJ, Miranda E, Alexander G, Huang-doran I et al (2010) Modeling inherited metabolic disorders of the liver using human induced pluripotent stem cells. J Clin Invest 120(9):3127–3136. doi:10.1172/JCI43122DS1

Sampaziotis F, Segeritz C-P, Vallier L (2015) Potential of human induced pluripotent stem cells in studies of liver disease. Hepatology 62(1):303–11

Sandborg RR, Smolen JE (1988) Biology of disease: early biochemical events in leukocyte activation. Lab Invest 59:300–320

Schwartz RE, Fleming HE, Khetani SR, Bhatia SN (2014) Pluripotent stem cell-derived hepatocyte-like cells. Biotechnol Adv 32(2):504–513. doi:10.1016/j.biotechadv.2014.01.003. Elsevier B.V

Scott CM, Kruse KB, Schmidt Z, Perlmutter DH, Mccracken AA, Brodsky JL (2007) ADD66, a, gene involved in the endoplasmic reticulum-associated degradation of alpha-1-antitrypsin-Z in yeast, facilitates proteasome activity and assembly. Mol Biol Cell 18:3776–3787. doi:10.1091/mbc.E07

Seersholm N, Kok-jensen A, Dirksen A (1994) Survival of patients with severe alpha-1-antitrypsin deficiency with special reference to non-index cases. Thorax 49:695–698

Serres F (2002) Worldwide racial and ethnic distribution of alpha-1-antitrypsin deficiency. Chest 122:1818–1829

Seymour C, Thomason M, Chalmers R, Addison G, Bain M, Cockburn F, Littlejohns P et al (1997) Newborn screening for inborn errors of metabolism: a systematic review. Health Technol Assess 1(11):1–95

Shapiro SD (2007) Transgenic and gene-targeted mice as models for chronic obstructive pulmonary disease. Eur Respir J 29(2):375–8

Sifers RN, Carlson JA, Clift SM, DeMayo FJ, Bullock DW, Woo SL (1987) Tissue specific expression of the human alpha-1-antitrypsin gene in transgenic mice. Nucleic Acids Res 15(4): 1459–75

Si-Tayeb K, Lemaigre P, Duncan SA (2010) Organogenesis and development of the liver. Dev Cell 18:175–189. doi:10.1016/j.devcel.2010.01.011

Stoller JK, Tomashefski J, Crystal RG, Arroliga A, Strange C, Killian DK, Schluchter MD et al (2005) Mortality in individuals with severe deficiency of alpha-1-antitrypsin. Chest 127:1196–1204

Sveger T (1976) Liver disease in alpha-1-antitrypsin deficiency detected by screening of 200,000 infants. N Engl J Med 294:1316–1321

Sveger T (1988) The natural history of liver disease in alpha 1-antitrypsin deficient children. Acta Paediatr Scand 77(6):847–851

Tafaleng EN, Chakraborty S, Han B, Hale P, Wu W, Soto-gutierrez A, Feghali-bostwick CA et al (2015) Induced pluripotent stem cells model personalized variations in liver disease resulting from a1-antitrypsin deficiency. Hepatology 00(00):1–11. doi:10.1002/hep.27753

Takahashi K, Yamanaka S (2006) Induction of pluripotent stem cells from mouse embryonic and adult fibroblast cultures by defined factors. Cell 126(4):663–676. doi:10.1016/j.cell.2006.07.024

Takahashi K, Tanabe K, Ohnuki M, Narita M, Ichisaka T, Tomoda K, Yamanaka S (2007) Induction of pluripotent stem cells from adult human fibroblasts by defined factors. Cell 131(5):861–872. doi:10.1016/j.cell.2007.11.019

Takayama K, Kawabata K, Nagamoto Y, Kishimoto K, Tashiro K, Mizuguchi H (2013) 3D spheroid culture of hESC/hiPSC-derived hepatocyte-like cells for drug toxicity testing. Biomaterials 34:1781–1789. doi:10.1016/j.biomaterials.2012.11.029. Elsevier Ltd

Tobin M, Cook P, Hutchison D (1983) Alpha 1 antitrypsin deficiency: the clinical and physiological features of pulmonary emphysema in subjects homozygous for Pi type Z. A survey by the British Thoracic Association. Br J Dis Chest 77(1):14–27

Touboul T, Hannan NR, Corbineau S, Martinez A, Martinet C, Branchereau S, Mainot S et al (2010) Generation of functional hepatocytes from human embryonic stem cells under chemically defined conditions that recapitulate liver development. Hepatology 51(5):1754–65

Travis J, Salvesen GS (1983) Human plasma proteinase inhibitors. Ann Rev Biochem 52:655–709
Werner ED, Brodsky JL, McCracken AA (1996) Proteasome-dependent endoplasmic reticulum-associated protein degradation: an unconventional route to a familiar fate. Proc Natl Acad Sci 93:13797–13801
Wernig M, Meissner A, Foreman R, Brambrink T, Ku M, Hochedlinger K, Bernstein BE et al (2007) In vitro reprogramming of fibroblasts into a pluripotent ES-cell-like state. Nature 448(7151):318–324. doi:10.1038/nature05944
Wilson AA, Ying L, Liesa M, Segeritz C-P, Mills JA, Shen SS, Jean J et al (2015) Emergence of a stage-dependent human liver disease signature with directed differentiation of alpha-1 antitrypsin-deficient iPS cells. Stem Cell Rep 4:873–885. doi:10.1016/j.stemcr.2015.02.021
Wu Y, Whitman INA, Molmenti E, Moore K, Hippenmeyert P, Perlmutter DH (1994) A lag in intracellular degradation of mutant alpha-1-antitrypsin correlates with the liver disease phenotype in homozygous PiZZ alpha-1-antitrypsin deficiency. Proc Natl Acad Sci 91:9014–9018
Yu J, Chau KF, Vodyanik MA, Jiang J, Jiang Y (2011) Efficient feeder-free episomal reprogramming with small molecules. PLoS One 6(3):1–10. doi:10.1371/journal.pone.0017557
Yusa K, Rashid ST, Strick-Marchand H, Varela I, Liu P-Q, Paschon DE, Miranda E et al (2011) Targeted gene correction of α1-antitrypsin deficiency in induced pluripotent stem cells. Nature 478(7369):391–394. doi:10.1038/nature10424

MIX
Papier aus verantwortungsvollen Quellen
Paper from responsible sources
FSC® C105338

If you have any concerns about our products,
you can contact us on
ProductSafety@springernature.com

In case Publisher is established outside the EU,
the EU authorized representative is:
Springer Nature Customer Service Center GmbH
Europaplatz 3, 69115 Heidelberg, Germany

Printed by Libri Plureos GmbH
in Hamburg, Germany